VIRUSES—THE INVISIBLE ENEMY

Dorothy H. Crawford is Emeritus Professor of Medical Microbiology at the University of Edinburgh. She has authored numerous scientific publications, and is the author of several popular science books including *Deadly Companions: How Microbes Shaped our History* (OUP, 2007, second edition 2018), *Viruses: A Very Short Introduction* (OUP, 2011, second edition 2018), *Virus Hunt: The Search for the Origin of HIV* (OUP, 2011), *Cancer Virus: The Story of Epstein-Barr Virus* (OUP, 2014), and *Ebola: Profile of a Killer Virus* (OUP, 2016). She is a Fellow of the Royal Society of Edinburgh, a Fellow of the Academy of Medical Sciences, and was awarded an OBE in 2005 for services to medicine and higher education.

VIRUSES –
THE
INVISIBLE
ENEMY

DOROTHY H. CRAWFORD

OXFORD
UNIVERSITY PRESS

OXFORD
UNIVERSITY PRESS

Great Clarendon Street, Oxford, OX2 6DP,
United Kingdom

Oxford University Press is a department of the University of Oxford.
It furthers the University's objective of excellence in research, scholarship,
and education by publishing worldwide. Oxford is a registered trade mark of
Oxford University Press in the UK and in certain other countries

First published 2000
First published in paperback 2002
Reprinted 2009
Second edition, as *Oxford Landmark Science*, published 2021

Impression: 1

Published in the United States of America by Oxford University Press
198 Madison Avenue, New York, NY 10016, United States of America

British Library Cataloguing in Publication Data
Data available

Library of Congress Control Number: 2021943061

ISBN 978-0-19-284503-0

Printed in Italy by
L.E.G.O. S.p.A. Lavis (TN)

This book is dedicated to
William, Danny, and Theo, and my extended family

VIRUSES: FOREWORD

When modern science began to take shape in Europe in the seventeenth century, and even as it developed in the eighteenth, the concept of a professional scientist was unknown. Those who actively investigated scientific matters, as well as those who were merely interested, were largely well-educated members of the upper classes. Indeed, this situation continued well into the nineteenth century, as evidenced by elections to the Royal Society of London, the UK's national academy of science and the longest-established scientific society in the world. During those centuries it was thus usual for well-informed people to be able to understand most of the science, or 'natural knowledge' as it was then called, which was being discovered and discussed at the time.

However, the latter half of the nineteenth century saw advances in science accelerate and the birth of whole new areas of research with their attendant academic disciplines; as a result, the pursuit of science became a career instead of a hobby. The past few decades have witnessed a veritable explosion in knowledge, with hundreds of new subjects of ever-greater complexity, huge and costly research undertakings, and a vast worldwide army of highly specialized scientists. The ordinary citizen, however generally well educated, has long been left behind in this unimaginably complicated era of new ideas—confusing, impenetrable, and unintelligible, except to experts.

This would not perhaps matter greatly were it not that the technological results of scientific progress impinge more and more on the way society functions and, even more important, on the way the individuals who make up society live their own lives. There now exist acute anxieties over what people perceive as the arcane and threatening progress of science, examples being the genetic manipulation of food and human reproduction, the polluting effects of technological advances on the environment, and the influence on behaviour of a globally penetrating mass media.

An obvious consequence of the public's bafflement over science is to be seen in the present anti-science backlash evident in Western countries, which is not only directed against science and scientists but also against the multinational companies employing the new technologies which advances in science have provided. It is regrettable that scientists have, on the whole, been reluctant to come forward and explain the nature of whatever highly specialized work they may be doing, and have likewise hesitated, particularly in the face of sustained tabloid hysteria, to stand up and make clear the huge benefits to mankind of recent scientific progress.

But without an understanding of what science is about and how it works, the public is not in a position to evaluate findings, make judgements about their impact on daily life, or appreciate the nature of risk or scientific uncertainty. It is therefore to be greatly welcomed that Professor Dorothy Crawford has now set out in this book a layman's straightforward and authoritative guide to virology, a subject constantly affecting our daily lives in both trivial and life-threatening ways. The extent of confusion and misinformation in this area is exemplified by the continuing inability of even the serious press to distinguish the fundamental differences between viruses and bacteria. Professor Crawford's reliable and readily comprehensible account of the viruses of man is thus more than timely, and clearly explains, in an engaging way, a field of science which is too often garbled by the uninformed. Her book will fascinate the curious public and should be obligatory reading for all those journalists who so regularly display their ignorance of this important subject.

Sir Anthony Epstein 2000

Professor Crawford's excellent review of the topic of virology should be obligatory reading for all those mystified by the COVID pandemic. The updated new edition contains a timely explanation of this novel, animal-derived virus disease which is so affecting all our lives.

Sir Anthony Epstein
Oxford, September 2020

PREFACE

December 2019—the month that changed our lives. COVID-19 arrived and suddenly no more hugs, kisses, or handshakes. No parties, meetings, or holidays. Schools, universities, restaurants, and hotels—all closed. Health services overwhelmed, COVID sufferers dying alone, and loved ones denied attendance at their funerals.

These are the conditions under which I am writing now, but when I wrote the first edition of *The Invisible Enemy* in 1999–2000, the HIV/AIDS pandemic was at its height. Scary, inflammatory, and frankly fictitious stories about HIV/AIDS appeared in the media, fuelling prejudices and engendering panic. Labelling the outbreak 'the gay plague', press reports encouraged discrimination with headlines like 'Place AIDS victims in quarantine', 'Blood on their hands', and even 'Exterminate gays'.

It seemed to me then that we were totally ill-equipped to deal with a pandemic, both mentally and physically. Before HIV, the assumption was that all severe infections could be either prevented by vaccination or treated with antibiotics. So the threat of infectious diseases was in general ignored as irrelevant, and investment in pandemic preparedness was way down governments' priority lists.

Consequently, when the HIV pandemic hit, people had little knowledge with which to logically assess the extreme views emanating from the press. Words like 'bugs' and 'germs' were familiar to everyone, and 'microbes' too in many cases; but few understood what viruses were, or how they spread, cause dis-ease, and kill.

So, as the twentieth century drew to a close, I sat down to write a book on viruses that was going to be at once accurate, informative, and entertaining. My passionate interest in, and admiration for, viruses evolved into a mission to write a book about them so fascinating that the general reader could not put it down!

I don't know if I succeeded in that mission, but now, at the beginning of 2021, the situation has completely changed. In the intervening 20 years, three severe epidemics/pandemics have hit us, all caused by viruses—SARS in 2003; swine flu in 2009; and, worst of all, COVID-19 from 2019—while the HIV/AIDS pandemic still continues to rage. Also, in the years immediately before COVID-19's emergence, there was an unprecedented number of emerging infectious disease outbreaks throughout the world, including the largest ever outbreak of Ebola in West Africa in 2014–16 and Zika virus in South America in 2016–17.

With life-threatening virus outbreaks, epidemics, and pandemics now so frequent, I see that people are better informed about viruses, and press reports more accurate, than they were 20 years ago. Also, there is a thirst for knowledge about virology, epidemiology, and infectious diseases. When I hear people on the bus having a casual conversation about the R number, I can be fairly sure that my pet subject is topical.

With this in mind, I have rewritten this history of viruses, taking the reader on a journey from ancient killers through to the modern age of rampant emerging viruses, providing the context for the emergence of SARS and COVID-19—caused by the two new killer coronaviruses. The book addresses how, where, and why new viruses emerge, and how we can prevent more surprise lethal viruses appearing.

I hope that the book provides interesting and thought-provoking reading.

D.H.C.

Edinburgh, January 2021

Viruses—The Invisible Enemy would not have been completed without the invaluable help and advice from Theo Alexander, William Alexander, Patricia Boyd, Richard Boyd, Kerrie Grant, Ingolfur Johannessen, Barbara Judge, Jenny Poller, Paul F. Saba, Jill Shepherd, Emma Stanton, and Mark Woolhouse. My heartfelt thanks to each of them. I am also indebted and Latha Menon and Jenny Nugée at Oxford University Press for their professional advice, patience, and encouragement.

CONTENTS

LIST OF ILLUSTRATIONS

Figures

Tables

1

THE VIROSPHERE

Viruses are everywhere, and with an estimated 100 million different types, there are more viruses on Earth than all other lifeforms put together. Every litre of sea water contains 10 billion of them! Viruses infect all living things, from single-celled bacteria to the largest and most sophisticated organisms. All this incredible abundance and diversity together is called the virosphere.

Viruses are obligate parasites

Viruses are unique. They are fundamentally different from all other organisms, including other microbes. They are smaller than the smallest of bacteria, and while bacteria and all other life forms are composed of cells—the basic building block of life—viruses are not (Table 1 and Figure 1). In fact, viruses are particles consisting of little more than a piece of genetic material protected by a protein coat. Or, put more crudely, 'a piece of bad news wrapped up in protein'.[1]

Viruses are honed to the bare essentials required to survive, albeit through a parasitic lifestyle. They cannot do anything on their own; so they are obliged to penetrate a living cell and

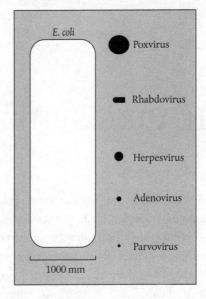

Figure 1. Comparative size of viruses and bacteria.

Table 1. Comparison of the characteristics of viruses and bacteria.

Criterion	Viruses	Bacteria
Size	20–300 nm*	0.5–5.0 mm*
Cell wall	No	Yes
Cell machinery for making proteins	No	Yes
Genetic material: DNA, and RNA	DNA or RNA	DNA
Reproduction	Only inside a living cell	Yes
Free living or parasitic	Obligate parasites	Mostly free living

*nm = nanometre
*mm = micrometre

take control. Viruses may appear to plan attack and survival strategies, but they are complete hostages to fortune. These so-called strategies are the outcome of millions of years of evolution, resulting in the survival of the fittest.

So what are viruses, and how can they be so dangerous?

Invisible invaders

At 20–300 nanometres long (a nanometre is a thousand millionth of a metre), viruses were completely invisible until the invention in the 1930s of the electron microscope that can magnify 100,000 times. Then the submicroscopic world of viruses was unveiled and their beautiful structures revealed.

Each virus particle simply consists of an outer protein shell, called a 'capsid', surrounding and protecting a valuable cargo— its genetic material. A virus's genetic material is either DNA or RNA, never both, and this tiny string of molecules bears its genes—the blueprint for future generations.

Most viruses have between 3 and 400 genes (compared to the human genome's 20,400 genes) that carry the code for their own reproduction. But in order to reproduce, these obligate parasites need access to vital chemicals, so they must invade a living cell to find them and access the cellular machinery.

Nowadays the term 'virus' is used to describe non-biological agents like computer viruses. And this is apt; these modern 'viruses' are invisible parasites that infect, reproduce, and cause disease, albeit in computer programs rather than living cells. Computer viruses are snippets of computer code embedded in emails, disks, or downloaded software. They only come to life and replicate once they have access to the computer hardware, which is analogous to the cellular machinery. And

DNA and RNA Viruses

DNA, or deoxyribonucleic acid, is a large molecule made up of two complementary chains coiled to form a double helix. Attached to each chain are the molecular building blocks called nucleotides that end in four 'bases' that form the genetic code (adenine, guanine, cytosine, and thymidine, usually abbreviated to A, G, C, and T). The chromosomes in the nucleus of a cell are composed of DNA that carries the genes for that particular organism.

The structure of RNA, or ribonucleic acid, is similar to DNA except that it is a single-stranded molecule and the nucleotide base thymidine is replaced by uracil (U). RNA is essential for downloading the genetic code and translating it into proteins. First the DNA sequence of a particular gene is used as a template for producing messenger RNA (mRNA). Then this RNA message conveys the genetic information to molecular machines called ribosomes, where proteins are assembled from strings of amino acids, each coded for by triplets of nucleotides.

Viruses may carry their genetic information as either DNA or RNA. DNA viruses (e.g. poxviruses, herpesviruses, and hepatitis B virus) tend to be larger than RNA viruses, while RNA viruses (e.g. measles virus, coronaviruses, and retroviruses) mutate more frequently than DNA viruses.

The retrovirus family, including human immunodeficiency virus (HIV), are so called because of their unique infection and survival strategy. Along with their RNA genome, retrovirus particles carry three enzymes, called reverse transcriptase, integrase, and protease. Once inside a cell, reverse transcriptase can reverse the normal flow of DNA → RNA to convert viral RNA into a DNA copy. Then the integrase enzyme integrates the newly made viral DNA into the host cell's DNA chain. From then on the viral DNA is treated by the cell as part of its own DNA chain and cannot be removed.

treatment packages are of limited use because viruses 'mutate' so frequently.

As long as a virus can get inside a cell, it can commandeer the cellular machinery that will read the virus's genetic code saying 'reproduce me' and get on with the job. So viruses turn cells into factories for virus production, and in a day or two, thousands of new viruses emerge. Infection often weakens or destroys cells, and if enough cells are affected it usually has a significant impact on the host. Whole organs can be wiped out, and if they are vital and irreplaceable then infection will be fatal. Rabies virus, for example, destroys brain cells, and Ebola virus kills cells which line blood vessels, causing catastrophic bleeding.

After infection, viruses behave in a variety of ways. Some, like the new severe acute respiratory syndrome coronaviruses (SARS-CoVs) and new flu viruses, spread rapidly between us to cause a dramatic epidemic or pandemic of an acute and deadly disease. But other viruses, like hepatitis and herpes viruses, are more subtle. They enter the body silently, then take up residence for the lifetime of the host, who is often unaware of their presence. These are the persistent viruses, some of which also cause epidemics and pandemics, and although these are less obvious and dramatic than the acute viruses, they can be just as deadly. For instance, persistent hepatitis viruses can cause death from liver failure. What's more, several persistent viruses, including some herpes and hepatitis viruses, cause cancer in a proportion of those infected. In fact, viruses are linked to around 15 per cent of all human cancers worldwide, including cancers of the liver, stomach, and uterine cervix, and also Hodgkin's lymphoma.

Weakening of the host cells that viruses inhabit can in some cases have unexpected and bizarre effects on the infected cells and on the animal or plant hosting the virus. One famous example of such an effect occurred in Holland in the seventeenth century, when beautiful variegated tulip flowers were first cultivated.

Tulipomania

Tulips were brought to Europe from Turkey in the mid-sixteenth century and Holland soon became the centre of the tulip trade. From a plain red flower, plant breeders developed variegated forms with white strips called 'colour breaks' (Figure 2). These

Figure 2. The variegated 'broken' tulip first cultivated in the seventeenth century.

so-called broken tulips were rare and much prized as a status symbol. But the flowers were as fickle as they were beautiful: 'The flower had a unique trick which added dangerously to its other attractions. It could change colour, seemingly at will.'[2]

Although once 'broken' a bulb remained so, only one or two in a whole field might show the effect, and none of the fascinated plant breeders could work out why. Those plants sporting flowers with colour breaks were less vigorous than their plain-coloured counterparts, and all this conspired to make the bulbs desirable and expensive. Between 1634 and 1637 a single prize 'Admiral van Enkhuijsen' bulb sold for 5,400 guilders (around £400,000 in today's money)—the cost of a town house in smart Amsterdam and the equivalent of 15 years' wages for a labourer. At the height of 'tulipomania' it was cheaper to commission a famous artist like Jan van Huysum to paint a picture of a 'broken' tulip (for up to 5,000 guilders) than to buy a single bulb.

And to think that this was all caused by a virus—a tiny cellular parasite carried by aphids from local fruit trees!

From miasma to germ theory

From the time of Hippocrates (460–375 BC) until the beginning of the nineteenth century, diseases were generally thought to be caused by two forms of poison: 'virus' and 'miasma'. 'Virus' referred to visible poisons like snake venom, the saliva of rabid dogs, and poisonous secretions of plants. 'Miasma' was an invisible gas that emanated from swamps, stagnant water, unburied human bodies, and animal carcasses, and caused infectious diseases and plagues.

Louis Pasteur in Paris and Robert Koch in Berlin, working in the middle of the 1800s, were the first to show that infectious diseases were caused by microbes. But it required an enormous leap of faith for a population not trained in scientific thinking to relinquish their old beliefs and accept that diseases were caused by tiny living things that colonized their bodies rather than by noxious substances coming from the outside. Not surprisingly, it took some time for the revolutionary 'germ theory' to be believed, and it was not until the middle of the nineteenth century that the theory was universally accepted.

By the beginning of the twentieth century several hundred disease-causing germs—that is, bacteria—had been identified, but there were still many diseases where no causative microbe could be found, including common, severe infections like measles, smallpox, rabies, and yellow fever, as well as foot and mouth disease of cattle, and tobacco mosaic disease of plants.

In 1876, Adolf Mayer, Director of the Agricultural Research Station at Wageningen, Holland, became probably the first to transmit a virus disease successfully. He worked on tobacco mosaic disease, so named because it causes the dark and light spots on the leaves of tobacco plants (Figure 3), and a disease of great economic importance to the Dutch as it could devastate their lucrative tobacco crops. Mayer found he could transmit the disease from one plant to another using mashed up leaves from infected plants, but he thought that the active ingredient must be an enzyme or toxin. Then another Dutchman, soil scientist Martinus Beijerinck, further unravelled the mystery by passing a diluted extract of infected leaves through porcelain filters with pores small enough to retain all known bacteria. The filtered extract not only remained infectious but also regained its original strength after passage to a second plant, so proving that the

Figure 3. Leaf of nicotiana (tobacco plant) showing mottled effect of tobacco mosaic virus infection.

disease-causing agent could reproduce and was therefore a microorganism of some sort.

There followed a period of intense and heated debate over these so-called invisible, or filterable, microbes, which were defined by three measurable characteristics: they were filterable—because of their small size they could pass through filters which retained bacteria; they were invisible—they could not be seen under a light microscope; and they were non-culturable—they would not grow on bacterial culture plates.

The term 'filterable virus' was coined to describe these microbes, regarded by many as merely very small bacteria. It was the invention of the electron microscope in 1938 that eventually resolved the issue by revealing clear images of viruses, so that at last the incredible structures and symmetries of these tiny particles could be studied in detail (Figure 4). Then, with the final cracking of the double helical structure of DNA by James Watson and Francis Crick in 1953, it eventually became clear that

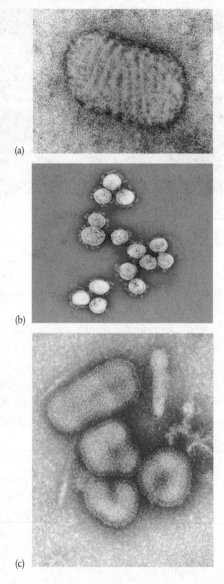

Figure 4(a)–(c). Electron micrographs of: (a) pox virus (orf) (×180,000); (b) SARS-CoV (×180,000); (c) influenza virus (×130,000).

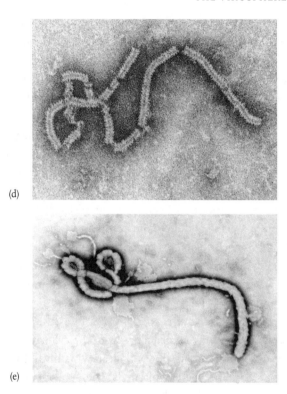

Figure 4(d)–(e). Electron micrographs of: (d) mumps virus (nucleo-protein helix, ×180,000); and (e) Ebola virus (×24,000).

viruses contained genetic material and the essential differences between viruses and bacteria were explained.

Where did viruses come from?

This is a difficult question to answer, because unfortunately viruses left no fossil records that might be used to track their past history. Opinions differ on their origin but all are agreed that although they were only visualized under a hundred years ago, they have been around for millions of years. Practically all living things, whether plant, animal, or microbe, carry their own particular viruses, which have evolved along with them from their earliest ancestors.

There are three theories on the origin of viruses: the progressive, the regressive, and the virus-first theories. Those backing the progressive theory believe that viruses arose from rogue pieces of DNA that can break free from chromosomes and rejoin in another position. Such segments are called transposons or 'jumping genes', and are quite common in the human genome. These genetic elements cannot move from cell to cell, but structurally they resemble retroviruses like human immunodeficiency virus (HIV), the cause of acquired immunodeficiency syndrome (AIDS). Retroviruses have an RNA genome that, once inside a cell, converts to DNA and joins the cell's DNA chain, thereafter being copied along with the cellular DNA.

One step further along the evolutionary pathway towards a virus may be represented by plasmids. These fragments of DNA reside inside bacterial cells but are not connected to the bacterium's chromosomes. They carry extra genes which sometimes confer new properties to the host bacterium. For instance, plasmids can provide the genes for antibiotic resistance or for harmful toxins. Like viruses, plasmids rely entirely on the cellular machinery for their reproduction, but they are not classified as viruses because they cannot exist outside the bacterial cell. Their only means of spread is during conjugation when two bacteria join briefly to swap genetic material, of which they may be part.

The regressive theory attacks the problem from the other end by suggesting that viruses may derive from bacteria that have regressed from a free-living to a parasitic lifestyle. This is supported by some of the largest viruses, such as the recently discovered mimivirus as well as poxviruses, that carry enzymes that make them less dependent on the host cell for their reproduction. At some point in the distant past, the theory goes, these were free-living organisms that developed a symbiotic relationship with their hosts, and eventually evolved into obligate parasites. Evidence for this also comes from certain bacteria, such as *Rickettsia* and *Chlamydia*, that live inside cells as obligate parasites.

The virus-first theory derives from the general acceptance of the recent hypothesis for the origin of life, in which the first self-replicating molecule to evolve was RNA, not DNA as previously assumed. If this is the case then RNA viruses could have evolved before the cells that they now depend on.

Individually these theories do not account for both RNA and DNA viruses, which may have evolved separately. So viruses could be descendants of living bacteria that have taken parasitism to its extreme, and/or pieces of cellular genetic material with a built-in code for their own reproduction, and/or remnants of the first life on Earth.

Alive or dead?

Viruses are generally talked about as if they are living organisms, but this assumption is up for debate. The *Oxford English Dictionary* defines 'life' as 'the ability to breathe, grow, reproduce, etc. that people, animals, and plants have before they die and that objects do not have'. This is rather unhelpful as far as viruses are concerned because they only become active after they penetrate a living cell. Although they possess some characteristics of living things, they lack many features essential for life. For instance, they can reproduce, albeit only with a great deal of help; but on the other hand, they lack all the metabolic processes needed to generate energy, and the molecular machinery required to make proteins.

Certainly, most molecular scientists would regard viruses as just another molecule of genetic material which can be manipulated in a test tube. Ultimately, does it really matter? Whichever way you look at it, there is no doubt that viruses are unique and utterly unlike the cells which form the building blocks of all other organisms.

How viruses spread

Every time a virus infects a host a battle ensues. A new strain of flu virus, for example, generally has just a few days to establish an infection before the host's immune response against it gets underway. During that time the flu virus must infect cells lining the respiratory tract and reproduce as quickly as possible. Then its offspring must exit the host before they are destroyed by its developing immunity. Whatever the outcome of the encounter for the host, sooner or later viruses must move on. Under normal conditions, viruses can survive from one or two days to a few weeks in the outside world depending on the type of virus and the situation it lands in. So after release from one host they must find and invade another rapidly if they are to maintain the unbroken chain of infection that guarantees their long-term survival.

Viruses do not take an active part in their spread between hosts because once outside a cell they are completely inert. They just have to take their chances in the outside world, drifting on air currents, floating in liquids, or lurking in food. Like the seeds of plants, many millions are produced, making it more likely that at least one will find and take root in a new host.

Viruses generally gain access to a host by infecting cells on body surfaces. They cannot penetrate our intact outer skin, but skin is contiguous with the inner epithelia, the layer of cells lining the surfaces of the intestinal, respiratory, and genitourinary tracts. So by entering through body orifices like the nose or mouth, they can find plenty of live cells to infect. Here they produce masses of viral particles which are ideally placed either to spread to internal organs or to hitch a ride in secretions or excretions back to the outside world.

Direct Spread

Some viruses spread directly from one person to another, and papilloma viruses are a good example. As children most of us have had harmless but unsightly warts on our hands or painful verrucae (plantar warts) on the soles of our feet. These are caused by benign papilloma viruses which penetrate the outer dead layers of our skin through a small cut or abrasion, to infect developing skin stem cells. To aid their own reproduction, these viruses stimulate infected cells to divide more rapidly than surrounding uninfected skin cells, so these virus-carrying cells push up to form the familiar tiny cauliflower-shaped lump of a wart, which in turn sheds more viruses. A mere handshake is enough to complete the cycle of infection as the virus jumps to another host.

Herpes simplex virus also infects developing skin cells, entering through a trivial lesion in the outer layers. This is generally on the face, usually somewhere near the lips. The infection causes cells to swell up, producing a painful rash of tiny blisters called a cold sore. Each blister is full of viruses and when they burst they release their cargo of new viruses, ready to spread the infection via a kiss.

In the genital area, herpes simplex virus type 2 causes genital herpes—the most common type of genital ulcer. This is also highly infectious and easily spread by unprotected sexual intercourse. But spread by sexual contact is not restricted to viruses which grow in the genital tract; viruses in the blood, such as hepatitis B and HIV, also take advantage of this route. Genital secretions usually contain a few white blood cells which may carry viruses, but also, just a minute amount of bleeding from injury or from genital ulcers is enough to increase the chance of sexual transmission of blood-borne viruses quite dramatically.

Food poisoning

Most of us have been mildly inconvenienced by 'tummy bugs', but these virus infections, causing gastroenteritis, can be serious. The viruses spread from person to person by the faecal–oral route, that is by ingestion of contaminated food and drinking water. They are adept at surviving passage through the acid environment of the stomach, and then growing in the small intestine. Here, they paralyse the superficial cells of the gut wall, causing acute diarrhoea and vomiting. Massive fluid loss results in dehydration, and it also flushes billions of viruses back into the environment, ready to attack their next victim.

Norovirus, the 'vomiting bug', is the most common cause of gastroenteritis, producing inflammation of the stomach and small intestine with consequent fluid loss. Symptoms appear after a short incubation period of just 1–2 days, with acute onset of stomach cramps, nausea, vomiting, diarrhoea, fever, and whole body aches heralding the infection. The result can be devastating, particularly in children and the elderly where rapid dehydration may be fatal. The virus survives in water for up to three weeks and just one is enough to seed the infection. Cruise ships are notorious hotspots for norovirus, where, in the enclosed, crowded environment of the ship, it can be very difficult to control—ruinous for idyllic holidays and ship company profits alike.

Rotaviruses are also highly contagious, infecting most children worldwide by the age of five. The virus thrives in the overcrowded and unhygienic conditions where most of the >500,000 annual deaths occur, mainly from dehydration. There are something like 1,000,000,000 (10^9) rotaviruses in every gram of faeces passed by an infected person. And, like noroviruses, these can survive for several weeks in sewage and water while waiting for an opportunity to infect a new host. A vaccine is now available to prevent rotavirus infection, and in recent years this has reduced childhood deaths considerably.

But food poisoning is not confined to developing countries and cruise ships. In 1977, when 120 eminent doctors gathered for their annual dinner at the Apothecaries Hall in London, they little thought what the consequences might be. Two to three weeks later, 50 guests, as well as the cook and a waiter, came down with jaundice. The epidemiologist on the case identified the culprit as hepatitis A virus, and the carrier was the dessert they had all eaten—raspberry parfait.[3] The raspberries in the parfait had been picked in Dundee, Scotland, two years previously, and frozen at a local factory. The source of infection was probably a single batch of about 20 contaminated punnets which were infected either by a factory worker or an unidentified picker in the fields.

This outbreak was bad enough, but during the investigation some startling practices among raspberry pickers came to light. Raspberries picked for freezing must be dry and perfect, so they are placed in small punnets. But raspberries picked for jam do not have to be in such prime condition, so they could be collected in large barrels. Pickers out in the fields developed the habit of urinating into these barrels because it increased the weight and thereby their pay. An unpleasant habit certainly, but fortunately not dangerous since any contaminating viruses would be killed by boiling the fruit during jam making. This revelation was instrumental in improving hand washing and toilet facilities at fruit-picking sites.

Coughs and sneezes

The old saying 'coughs and sneezes spread diseases' is quite correct; a cough produces airborne mucous droplets, while a sneeze generates a much finer spray of fluid secretions, like an aerosol from a pressurized can. Viruses which cause colds and flu grow in the nose, throat, and sinuses, where they stimulate the sneeze reflex, causing millions of virus-laden droplets to be released into the air from an infected person, ready to be inhaled by cohabitants of a crowded room, train, bus, or plane. This is the most common and efficient method of virus spread in temperate climates, and is the route used by the emerging SARS-CoVs to amazing effect. Both SARS and COVID-19 are caused by airborne viruses that infect the respiratory tract and can cause pneumonia. But there is an important difference in their ability to jump between hosts that in part accounts for our differing success in controlling their spread. SARS virus particles are mainly produced from cells lining the lung cavities and are projected into the air in relatively heavy mucous droplets. These fall to the ground fairly rapidly and therefore do not travel far from the patient. Thus transmission is generally restricted to close contacts such as medical care staff and family members. In contrast, the virus behind the COVID-19 pandemic is spread in a fine spray of lighter droplets that remain suspended in the air for longer and travel farther from the victim. This difference is enough to make COVID-19 far more challenging to control (see Chapters 3, 4, and 5).

Worldwide, the common cold is the most frequent virus infection. Most of us catch a cold about twice a year, suffering the familiar symptoms—runny nose, swollen and watery eyes, and a general feeling of congestion. Although apparently trivial, colds have major economic significance, being one of the commonest causes of absence from work.

In post-Second World War London, the medical virologist Sir Christopher Andrews was on a mission to identify the elusive common cold virus. He needed an isolation unit for the task, and just outside Salisbury, Wiltshire, he found what he was looking for. From a wartime American hospital built in 1941 and intended to house victims of the epidemics they confidently expected as a result of war-time bombing, he constructed his Common Cold Research Unit.

The Unit started operations in late 1946, and for several decades generations of volunteers spent ten days in fairly basic accommodation, with only a little pocket money and a 40 per cent chance of catching a cold. Amazingly, there was no shortage of applicants—in all, 11,000 human guinea pigs passed through the doors, including holidaymakers, honeymoon couples, and some who apparently enjoyed the experience so much that they came back again and again. Fifteen visits was the record.

Inmates were housed in pairs, and quarantine rules were strict. They were allowed out of their rooms but had to observe the '30-foot rule'—that is, they were not allowed to approach within 30 feet (10 metres) of another human being. In his book entitled *In Pursuit of the Common Cold*, Andrews notes that:

> On very infrequent occasions volunteers have flagrantly and
> repeatedly ignored the rules, as when members of opposite
> sexes have associated together: and then it has been necessary
> to dispatch them home forthwith, withhold their pocket
> money and exclude them from the trial.[4]

If no colds developed after five days' isolation, volunteers were
given nasal drops of concoctions from the laboratory containing either putative cold viruses or a control fluid. Thereafter the
daily inspection of paper hankies became an important ritual.
Once used, these soggy items were stored as vital evidence of a
cold. Five days later volunteers were free to go, presumably
feeling that they had done their bit for science.

In the laboratories attached to the Unit, scientists were making desperate attempts to isolate the culprit virus. First it had to
be grown in culture, but when growth on egg membranes,
embryonic tissues, and many other media failed to entice to
the virus to grow, the scientists turned to experimental animals.
Ferrets, mice, rats, hamsters, guinea pigs, rabbits, squirrels (both
grey and flying varieties), voles, hedgehogs, kittens, pigs, monkeys, and baboons were all tested; but all proved resistant.

It was not until the British virologist David Tyrrell joined the
Unit in 1957 that the breakthrough came. He advised lowering
the temperature of the cell cultures from body temperature
(37°C) to 33°C, the temperature of the nasal passages where the
virus grows naturally. At last a virus grew! And when tested on
volunteers it produced the symptoms of a cold. This virus was
officially named rhinovirus in 1963, from the Greek word 'rhis'
meaning 'of the nose'.

We now know not only that there are 100–150 different types
of rhinovirus, but that the common cold can also be caused by
many other viruses, each of which have multiple strains or subtypes, so that there are in reality several hundred 'cold' viruses.

In 1965, Tyrrell discovered the first human coronavirus and, with electron microscopist June Almeida, described the virus particle and gave it its name. There are four known strains of common-cold-causing human coronaviruses, but they prompted little interest until the emergence of their more aggressive cousins, the SARS viruses, 40–50 years later (see Chapter 5).

Hitching a ride

Microbes may be carried from host to host by many different insect vectors, the best known being the unicellular malaria parasite *Plasmodium*, transmitted by the *Anopheles* mosquito. Several viruses, the first identified being that causing yellow fever, also exploit blood-sucking insects such as sand flies or mosquitoes to transport them between hosts. Animals generally act as a natural reservoir of the virus, which is spread between them by mosquitoes in a never-ending animal–insect–animal cycle. Any human who happens to encroach on this cycle by being bitten by a virus-laden insect may get infected. Yellow fever, dengue fever, and Zika viruses are all insect-borne emerging viruses causing problems in humans. They all naturally infect monkeys, and are transmitted between them by female mosquitoes that require a blood meal to nourish their developing eggs. So if a human is bitten by a virus-carrying mosquito this can cause disease that can also be passed on to others by these insects. Because the viruses' life cycles are so intimately linked to that of the female mosquito, wherever she thrives so can the viruses. And as climate change is increasing the range of tropical mosquitoes, so these diseases are spreading to new territories. They are now classed as emerging infections and are discussed in full in Chapter 3.

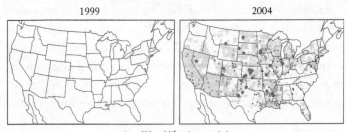

Any West Nile virus activity
Incidence per million in humans : · 0.01–9.99 • 10.00–99.99 ● ≥ 100.00

Figure 5. Maps showing the spread of West Nile fever virus across the US.

Another mosquito-transmitted virus is West Nile Fever virus, which suddenly turned up in New York in 1999 (Figure 5).[5] This virus is endemic to Africa, Asia, parts of Europe, and Australia, where its primary hosts are birds. It is spread among them by mosquitoes, and may spill over to humans if they are bitten by a virus-carrying mosquito, but it does not transmit between humans. Infection is usually asymptomatic or manifest as a mild flu-like illness, but inflammation of the brain (encephalitis) occurs in around one in 150 cases. How this virus reached the US is a mystery, but by exploiting the virgin US bird population and local mosquitoes, West Nile Fever virus spread in a wave across the US, reaching the West Coast, Mexico, and the Caribbean in just four years. And the virus is there to stay, mainly residing in the western states. In 2020 there were a reported 557 cases, with many more mild and asymptomatic infections.

Man-made virus transmission

Viruses have been quick to exploit modern medical practices. They can find their way into new hosts in transfused blood,

blood products, and organs for transplant, and can contaminate injecting equipment, surgeons' knives, and dentists' drills. The main culprits are blood-borne viruses such as hepatitis B and C and HIV, which all establish persistent infections in apparently healthy people who are often unaware of their presence. Of the three, hepatitis B is the most infectious; each millilitre of the blood from a carrier may contain several million viruses, so infection can be transmitted by a microscopic amount.

These days, blood donors are carefully selected and donated blood is screened using highly sensitive tests, so the chance of catching any virus from a transfusion is extremely small. Less than one in a million units of blood is contaminated in western countries, although this figure is higher in countries where screening is not as rigorous and/or the incidence of these persistent infections is higher.

Clearly, people who need regular infusions, such as haemophiliacs requiring blood-clotting factor VIII, are more at risk of blood-borne viruses. In the 1980s, just after HIV was discovered, contamination of donated blood caused an HIV epidemic among haemophiliacs in the US and Europe (see Chapter 4). Similarly, in Africa the spread of HIV was probably facilitated by the reuse of non-sterile needles, particularly during the large smallpox vaccination campaigns in the 1970s.

Persistent hepatitis C virus infection is most often acquired through the use of unsterile injecting equipment, and so is common among intravenous drug users. Like hepatitis B, persistent infection with hepatitis C virus causes liver damage which may lead to cirrhosis and/or cancer (see Chapters 6 and 7). This is an ongoing public health problem, particularly in Eastern and Southern Europe and the Eastern Mediterranean regions, where many are unaware of their infection. But the highest world incidence is in Egypt, where the virus was spread unwittingly during a 1960s treatment programme for the parasitic disease Bilharzia when unsterile needles were used.

A well-documented outbreak of hepatitis C virus in Valencia, Spain, which infected 217 people, was eventually traced to an unusual source.[6] All those infected had undergone a surgical operation in one of two local hospitals, the state-run La Fe or the private clinic Casa de Salud, in the previous two years. The link turned out to be the anaesthetist, who was not only infected with hepatitis C but was also a long-standing morphine addict. When administering morphine as a painkiller immediately after an operation, he would conveniently help himself to a shot before administering the remainder to the patient using the same needle, now contaminated with his hepatitis C-infected blood.

The most extraordinary case of man-made virus transmission I ever heard of is that of rabies virus transmitted by organ transplant. This virus, very rare in the West, is generally transmitted through a bite from a rabid animal, and infection is 100 per cent lethal (see Chapter 3). But the virus was reported in Germany in six transplant recipients who had become infected because the organ donor had, unbeknown to all, died of rabies. Of the six recipients, three died of rabies and two who received corneal grafts survived, as did the recipient of the liver who had previously been vaccinated against rabies.

Mother-to-child transmission

The Australian ophthalmologist Sir Norman Gregg, working at The Royal Alexandra Hospital in Sydney, was the first to discover that viruses could spread directly from mother to unborn child. In 1941 he noticed an unusually high number of babies born with cataracts and heart abnormalities, and he linked this to German measles (rubella) infection in the mothers who had been pregnant during an epidemic of the disease the previous year.[7] Viruses in the mother's blood can cross the placenta, grow in the baby, and damage developing organs. The earlier in pregnancy that German measles occurs, the higher the risk to the baby, and the more severe the damage. Infection during the first month of pregnancy, when the baby's organs are developing, affects almost all babies, whereas after the fourth month, when the baby is fully formed, no abnormalities occur. Of course only those mothers who had escaped German measles as a child were at risk of catching it during pregnancy, and nowadays the risk can be eliminated by childhood rubella vaccination.

As well as spreading directly from mother to child through the placenta before birth, persistent viruses like hepatitis B in the blood and herpes simplex in the birth canal can infect a baby during birth, while others, such as cytomegalovirus and HIV, cross from mother to child in breast milk (see Chapter 6).

In this chapter, we have discussed what viruses are and how they jump between hosts. But finding a new and susceptible host is only the beginning of the virus life cycle that ends in the production of new virus offspring. In the next chapter, we look at some of the barriers viruses must overcome in order to complete this cycle.

2

THE FIGHTBACK

In order to survive, viruses must penetrate host cells before they can begin the process of reproducing their genetic material, and here again viruses appear remarkably resourceful. Normal cells are bathed in a sea of chemicals, like hormones and growth factors, all looking for a way to interact with and influence the cell in some way. But this interaction is restricted by a series of receptor molecules on the cell surface that act as locks, each of which can only be opened by the particular chemical whose molecular key fits it precisely. This ensures that each type of cell behaves appropriately—that only nerve cells respond to nerve cell growth factor, T cells to T-cell growth factor, and so on. But this mechanism also provides viruses with a port of entry. By carrying a molecular key on their surface, they can disguise themselves as normal body constituents, latch on to, and enter, any cell that bears the complementary lock (Figure 6).

So viruses infect only those cells that display the particular molecular lock that their key fits into, and this restriction dictates the type of cell a virus infects and therefore the symptoms it will cause. Since there are several hundred molecules to choose from, viruses cause a great variety of diseases. A well-known example is HIV, which carries the key for a cellular

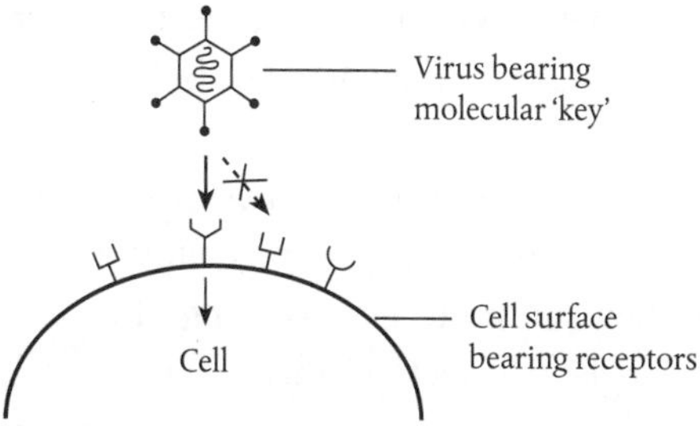

Figure 6. Virus infection of a cell. A virus uses its 'key' to enter a cell via a particular receptor molecule on the cell surface.

molecule called CD4 and therefore infects and destroys CD4-positive cells. These cells are central to the functioning of the immune system and, as we will discuss in Chapter 5, when they are destroyed by HIV infection faster than they can be replaced, the body's immune defences crash and AIDS is the result.

So far we have looked at infection from the viruses' point of view—how they invade a host and penetrate and commandeer cells for their own ends. The apparent ingenuity viruses display is astonishing, but they are not fighting a one-sided battle. Even the simplest organisms have ways of dealing with viruses, but the sophistication and subtlety of the human immune system is unrivalled.

As soon as we are born, and every day of our lives, our bodies are like castles under siege, surrounded by enemy troops all trying to breach the walls, enter, and plunder the contents. Each enemy carries a weapon to assist it and each tries a different port of entry. But just like fortified castles, we are built to withstand the attack.

The first strategy is to prevent access. We are covered with a thick layer of skin which, when intact, is impenetrable. It comprises several layers of brick-like cells all linked together and covered with an outer layer of flattened dead cells. Viruses cannot infect dead cells because they are metabolically inert, so they must enter by injection or injury, or through a natural orifice.

Although our outer skin is continuous with the cells lining the gastrointestinal, respiratory, and genitourinary tracts, these inner linings have no protective covering of dead cells and are often just a single cell thick. Here viruses (and other harmful microbes) can get a foothold, but we can often forestall this line of attack. Secretions like tears and mucus, as well as urine, all contain antiseptic substances, while the vagina and stomach contain acid that destroys all but the hardiest of attackers. Cells lining the upper respiratory tract bear small hairs, called cilia, which beat in unison, acting like an escalator to carry foreign particles up and out. Invaders that successfully bypass all these traps are then gobbled up by specialized cells called macrophages (meaning 'large appetite') that tour the tissues of the body, engulfing and destroying foreign particles. Once they have devoured an invader, macrophages release a variety of chemical signals called cytokines, which increase blood flow to the area and send immune cells, in the form of B and T cells (lymphocytes), scurrying to the scene.

Like red blood cells carrying oxygen to tissues, B and T cells patrol all the far-flung corners of the body by travelling in the blood along arteries and veins. But they also gather in lymph glands, where they communicate with each other. Lymph glands are strategically placed to guard danger zones where microbes are most likely to attack. The tonsils and adenoids protect the entrances to the lungs and the intestine, the walls of the small and large intestines are rich in lymph glands, while glands in the groin and armpits provide a stockpile of immune cells to guard the limbs. Macrophages from sites of infection gravitate to lymph glands, where they pass on vital information about foreign invaders to the resident lymphocytes.

B and T cells play vital roles in the body's defences. Their importance is amply demonstrated by rare genetic accidents in which one or other is absent or non-functional. Babies born without B cells cannot make antibodies. These children have no particular trouble combating viruses but have major problems with bacterial infections unless they are given regular antibody infusions. In complete contrast, babies born with no T cells have no trouble with bacteria, but suffer devastating virus infections, and cannot survive long without a bone marrow transplant to correct the defect. This natural accident tells us that T lymphocytes are vital for protection against viruses. Antibodies directed against the virus receptor, called neutralizing antibodies, can control the spread of viruses in the body, but it is those T cells known as 'killer cells' which are essential for eliminating the virus, because they search out and destroy virus-infected cells before they have time to unload their cargo of new viruses.

It is interesting to compare a microbial infection with an acute poisoning. An overdose of aspirin, for example, causes a metabolic upset that continues until the drug is neutralized by the liver. Overdoses of the same drug taken a week later, and again the week after that, will have the same effects. In fact, however many times you overdose on aspirin your body will never get used to it. But this is not the case with infections because of immunological memory. After you have been infected by, say, the measles virus, you will be immune to the virus for life. And although immunity to certain viruses fades with time, infections such as flu and the common cold, which seem to recur with monotonous regularity, are sometimes caused either by different viruses that give similar symptoms, or by the same virus that has mutated to escape immunological memory.

To cope with the huge number of different microbes, the body produces an estimated 50,000,000,000 (5×10^{10}) B and T cells every day. Each lymphocyte bears a receptor that can only recognize one small section of a foreign protein, called a peptide. Most are destined to lead a short and useless life, dying without ever meeting their particular peptide. But if one B or T cell does happen to meet its specific peptide, from, say, a flu virus protein on the surface of an infected cell, then that particular lymphocyte is galvanized into action. It divides rapidly to form a clone of identical daughter cells. The cloned B cells

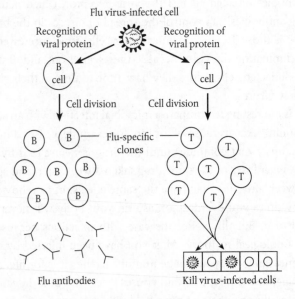

Figure 7. Clonal expansion of B and T lymphocytes. Contact with a virus causes specific B and T cells to multiply. B cells produce antibodies against the virus, whereas T cells can kill virus-infected cells.

produce antibodies that bind to the flu virus peptide for which they bear the receptor, while T-cell clones kill any flu virus-infected cells they come across (Figure 7). The first time you are infected by a virus it takes five to ten days to generate these clones, so the virus has time to multiply sufficiently to cause symptoms. But clones of B and T cells, once established, are retained and are ready to react quickly and prevent disease the next time the same virus comes along, and that is the basis of immunological memory.

The arms race

Clearly the human body is a battleground, with viruses attacking and the immune system defending on a regular basis. Thankfully, in most cases our immune system wins out. But now we need to assess how this daily battle influences each of the combatants in the longer term. This means looking back over millions of years to see how viruses and their hosts have co-evolved.

The evolution of all microbes and the diseases they cause has paralleled, and been shaped by, development of efficient ways of preventing and fighting them in their respective hosts. So as the army of microbes has evolved more and more ingenious methods of attack, the human body has retaliated with further refinements to its defences, so raising the struggle for survival to an unbelievably sophisticated level.

The viruses of the ancient herpes family are incredibly wide-spread in nature—even primitive invertebrates have their own strains. These and other viruses have diverged and evolved with their hosts over a similar time frame. So when mammals diverged from their reptile ancestors some 220 million years ago, herpes viruses were already in residence and diverged with their respective hosts. Then, around 80 million years ago, the evolution of mammals accelerated towards today's modern species, and contemporary herpes viruses followed their lead. Hence, human herpes viruses are most closely related to the herpes viruses that infect our closest relatives—Asian and African primates—and they have diverged from each other to the same extent as their individual primate hosts have.

On an evolutionary timescale, the genes of both host and virus have sustained many mutations. This may alter the nature of a protein, and although most mutations are either neutral or harmful, the occasional change is advantageous, and then the mutated organism becomes more successful than its rivals and takes over the population. This is the basis of coevolution, which, over millennia, has so exquisitely tuned viruses to their respective hosts that, as we noted earlier, it often looks as if their survival strategies involve detailed planning rather than random trial and error.

In the short term, viruses generally have the advantage over their hosts because they have a much shorter generation time. Viruses produce thousands of offspring every one to two days, whereas in most developed countries, an individual human has an average of 1.5 descendants in 20–30 years. And since each round of virus replication throws up several mutated offspring, clearly more complex organisms will take longer to adapt to a virus than vice versa. In fact, it is estimated that it would take humans 120–150 years (4–5 generations) to adapt to a new killer virus.

A famous illustration of adaptation by coevolution comes from the man-made epidemic of myxomatosis in rabbits. In the 1850s, European rabbits were introduced into Australia for hunting and as a source of food. But with no natural predators their numbers escalated rapidly until there were an estimated

300 million rabbits causing around £300 million worth of damage to Australian crops every year. What's more, they ate native plants and displaced natural wildlife species.

So with farmers clamouring for action, the Australian Government decided to release rabbit myxoma virus, a relative of smallpox virus, which naturally infects Brazilian rabbits. This virus is spread between the Brazilian rabbits by biting and blood-sucking insects, and causes them very little harm. However, it is lethal to Australian (European) rabbits, and initially, it had the desired devastating effect. In the first year, it killed 99.8 per cent of infected rabbits and the population reached rock bottom three years after the release of myxoma virus. But this high death rate soon began to drop, and seven years later only 25 per cent of infected rabbits were dying. Eventually rabbits were back to full strength in Australia and another virus was recruited to control them—rabbit haemorrhagic disease virus, a lethal and contagious virus for European rabbits.

Rabbit haemorrhagic virus is spread by many routes including by direct contact, in airborne droplets, and by insect vectors, so its impact is far-reaching. In 1995, this virus was released under strict quarantine conditions at a trial site on uninhabited Wardang Island, 5 kilometres off the coast of South Australia. But somehow, probably as a result of a forest fire driving virus-carrying mosquitoes off the island, it escaped to the mainland. On reaching Flinders Range National Park, the virus killed 750,000 rabbits without affecting humans or other wildlife in the area, so scientists then released it in hundreds of locations around the country. In most of these release sites, the rabbit population dropped by 95 per cent, and, in contrast to myxoma virus, this lethal outcome was sustained for many years, although recent reports suggest that now, 25 years later, rabbit numbers are beginning to rise in some areas.

Using stored material from before and after myxoma virus release, scientists have uncovered the genetic basis for the rapid change from lethal to non-lethal infection in Australian rabbits. Apparently, the rabbits adapted to the virus by the mutation of many genes known to influence the immune response to viruses. One particular mutation increased the effects of interferon, a protein with a strong anti-viral action. At the same time, the virus mutated to become less deadly, and in the end viruses with moderate virulence became dominant in the field.[1]

It is often assumed that coevolution of virus and host invariably leads to milder disease over time since it is usually to the advantage of both parties, but this is not always the case. Although we do not know the genetic details of the rabbit haemorrhagic fever virus–host interactions, in certain places this virus actually increased in virulence over time. This was due to a combination of factors, one being that while the disease was near 100 per cent lethal in mature animals, young rabbits were resistant to lethal disease. Thus the rabbit population was maintained despite the virulence of the virus in older animals, and so there was no selection pressure on the virus to make it less virulent. Also, the virus is mostly spread by insects, such as flies that feed on rabbit carcasses, so the more viruses there are by the time the host dies, the greater the spread of the virus by insects. This means that there is selection pressure on the virus to produce many offspring as rapidly as possible, and this trade-off would maintain its virulence.[2]

Some virus infections cause no disease at all, while others kill rapidly; but between these two extremes there is a spectrum of possibilities. The best strategy for each virus depends on its method of spread, incubation period, site of infection, time taken to reproduce, and the symptoms it causes. So, for example, a virus spread by aerosol, like the common cold virus, needs to be inhaled quickly by a new host before its transporting droplet dries out. Therefore, it spreads best in crowded conditions. As it only causes mild symptoms of rapid onset and short duration those infected do not generally retire to bed but continue to meet, and inadvertently infect, other people during the short time available between infection and the immune response putting a stop to production of new viruses.

By contrast, yellow fever virus employs mosquitoes to carry it from one victim to another, and relying on an insect vector is a precarious existence for a virus. So, on the basis that the longer the insect feeds the more likely it is to pick up a large dose of virus, the best strategy for this virus is to render its victim too ill to move around or bat away a feeding insect. Here, host isolation is not a problem for virus transmission because a mosquito keeps the virus alive for several days while transporting it anywhere within its flight range of up to 2 kilometres.

Viruses that spread by sexual contact exploit the basic requirement for survival of a species. In doing so you might imagine that they cannot fail to survive themselves. However, this is not an efficient method of spread in societies in which people are either monogamous or have few sexual partners over a lifetime. In the extreme case, one monogamous, infected individual will pass the virus to their partner, and there the trail will end. So there is pressure on the virus to increase its potential for spread in order to maximize its chances of survival. Most viruses that are spread by sexual contact, like herpes simplex virus type 2, manage this very successfully by evading the immune response and remaining in the body for long periods as a silent infection. Periodic bursts of new virus production over the lifetime of the host then allow this virus the greatest potential for spread to new sexual partners.

Our most recently acquired sexually transmitted virus, HIV, maximizes its chances of success in a similar way. It establishes a silent infection during which the person remains well and sexually active, but also *infectious*, for many years. Although HIV may become less virulent with time, this virus evolves rapidly and virulence might still get the upper hand. This balance depends on the average number of sexual partners per person in a community. In theory, if this average number increased, then the virus could spread more easily and quickly among community members. This being the case, a more aggressive strain of HIV, with a higher rate of virus production (and consequently more rapidly fatal disease), would be more successful than the slower-growing form, and would then become the dominant strain.

Viruses like HIV and herpes simplex, which establish chronic infections and in effect play hide-and-seek with immune cells, are discussed in more detail in Chapter 6.

What threats do viruses pose?

This is a difficult question to answer. When HIV was first discovered in the 1980s, it was suggested by some hyped-up press reports that it could exterminate the human race. This was primarily because it killed virtually 100 per cent of those it infected, albeit after a latent period of around 10 years. But although no effective treatment was available at the time, and even now no vaccine has been developed, effective antiviral drugs were eventually generated and now those living with HIV have a normal life expectancy as long as they take the medication. Also, we now know that around 1 per cent of Caucasians are resistant to HIV infection. This turns out to be due to a mutation in part of the cell receptor for HIV, a molecule called CCR5, that is required along with CD4 for the virus to bind to and enter a cell. This genetic variant may have protected carriers from a lethal infectious disease many hundreds of years ago, perhaps the bubonic plague (caused by the bacterium *Yersinia pestis*) in the Old World (1346–1666). This would cause the mutation to increase in incidence in the population, eventually reaching around 10 per cent. But only those who are homozygous for the mutation—meaning that they have inherited it from both parents (around 1 per cent of the population)—are protected from HIV infection.

This finding emphasizes the power of natural selection, and also the vast genetic variation seen between individual humans that determines our response to infections. Thus it is likely that there will always be variation in the severity of illness between individuals suffering from the same disease, and a proportion will survive even the most deadly of viruses.

One threat still taken seriously by governments is the use of viruses (and other microorganisms) in biological warfare. This type of warfare is not by any means new, and now that we have access to state-of-the-art delivery devices that can disseminate any lethal microorganism widely and efficiently, it is of international concern.

From time to time throughout history, people and governments around the world have used microorganisms as an efficient and cost-effective weapon of mass destruction. Starting in a rather crude way, the Greeks and Romans deposited dead animals into their enemies' drinking water. Later, dead soldiers were used in this way, and the technique was further refined in medieval times when the bodies of people who had died of infections were catapulted into besieged towns. In 1763, in the earliest recorded deliberate release of a virus, Sir Jeffery Amherst,

British Commander-in-Chief in North America, authorized the distribution of smallpox-contaminated blankets to native Americans who were harassing European settlers around the garrison at Fort Pitt in Pennsylvania.[3]

These days, the main problem is likely to come from terrorist groups or megalomaniac dictators, and we have recently seen both employing chemical poisons to target their perceived enemies. However, biological weapons have certain advantages over chemical poisons as they are cheaper and relatively easy to prepare. Although restrictions are in place, seed cultures of many dangerous microorganisms can still be obtained from national collections, and since making vaccines is a legitimate reason for growing microbes on a large scale, biological weapons factories can masquerade as vaccine-production plants.

Microorganisms can be smuggled through traditional security devices, and only tiny volumes are needed to kill huge numbers of people. Furthermore, since they are invisible, odourless, tasteless, and have a delayed action, they can be released into the air without immediate detection. Finally, as we have no previous experience of this type of attack, they are bound to unleash such panic and psychological trauma that total confusion would reign. However, in today's interconnected world, this strategy could be risky as, unlike chemical poisons, microbes can multiply and spread internationally, perhaps endangering those who released them and their allies.

Many different organisms have been tested for their potential as agents of biological warfare, including the bacteria that cause TB, typhoid, plague, cholera, botulism, anthrax, and gas gangrene. Candidate viruses include rotavirus that produces incapacitating vomiting and diarrhoea, and haemorrhagic fever viruses like Ebola, which is up to 90 per cent lethal. But the most effective biological weapons are now considered to be the bacteria causing anthrax, used in an attack in the US in 2001, and smallpox virus.

Because smallpox was eliminated from the natural environment in 1980, vaccination programmes have ceased and the world's population is once more susceptible. The virus causes a devastating and often lethal disease, which spreads rapidly in large and crowded cities. But best of all from the point of view of the aggressor, the virus remains infectious for long periods, so that it could be packed into the warheads of guided missiles and sent to its destination still in a viable condition. The threat of a deliberate smallpox release is well recognized by governments, and stocks of vaccine have been retained for this eventuality. Preparations for 'Operation Desert Storm' during the Gulf War of 1990 included vaccination of US and British troops against smallpox, but in reality it would be impossible to vaccinate an entire population in time to prevent an epidemic. However, biological weapons are primarily designed to destroy all vital activity but not necessarily to wipe out the whole human race. And smallpox, although it would certainly incapacitate, would not kill all those infected, not least because of our inbuilt resistance, developed and strengthened during the centuries when the virus was rife.

In this chapter we have seen the ingenious ways in which viruses sneak into our bodies and hijack our cells. Then, by adapting rapidly to any changing situation, they can outwit our immune systems and gain the advantage—in the short term at least. Such adaptability confers unpredictability, and this engenders fear and panic. Concern over what new and emerging viral threats may be in wait for us is constant and real, and this topic, including the COVID-19 pandemic, is discussed in Chapters 3, 4, and 5.

3

EMERGING
INFECTIONS

The appearance of a new virus infection, a so-called emerging infection, is not just a chance event. There are always rational answers to the questions 'why?', 'how?', 'when?', and 'where?' With every new infectious disease, answers to these questions are diligently sought by teams of epidemiologists dedicated to chasing up reports of strange new illnesses anywhere around the world. Once the culprit is identified, scientists map its genes to discover which virus family it belongs to. Then they plot an evolutionary tree, identify its closest relatives, and deduce when the 'new' virus diverged from its ancestors and its likely source.

Strictly speaking, emerging viruses are not entirely new. They have usually been around for millions of years but have come to our notice because they have changed their habits in some way. Most often they are zoonotic viruses, that is, viruses that jump to humans from the animals they naturally infect—their primary hosts. Viruses and their primary hosts have generally coevolved for so long that the virus is relatively harmless to its host. Problems only arise when for some reason a virus crosses a species barrier and colonizes a new host. Occasionally it is a genetic mutation which allows this to happen, but this is much less common than was once thought. Mostly it is our close contact with the primary hosts that is the all-important event.

Nowadays, in the Anthropocene, human activity is having a significant impact on the Earth's ecosystems, and many factors related to these changes interact to produce the ideal circumstances for a new virus to jump to, and flourish in, humans. We have an ever-growing world population, and this inevitably means that we are impinging on the natural environment more than ever before. Colonization of new territories, deforestation for development of new farming land, intensive factory farming, large irrigation schemes, wars, and climate change are just a few of the ways that we encourage emerging viruses. The increased demand for wild and exotic species for food or as pets adds to the danger. African bushmeat, for example, now has international appeal as a luxury food item; yet, as we will see in Chapter 4, it has been responsible for transmission of deadly viruses. Even more hazardous are the traditional East Asian live animal markets, or so-called wet markets, where wild animals are sold live for the table. These markets are undoubtedly hotspots for virus swapping. They offer a large variety of wild animal species, either hunted or bred in captivity, including rare species such as the critically endangered pangolin. This variety, combined with the cramped, unhygienic conditions of their storage, allows viruses to cross between species; occasionally onward passage to humans occurs, and as these markets are generally situated in towns and cities, viruses have immediate access to large populations of susceptible new hosts to infect.

Presently, new human viruses are emerging at an average rate of two to three per year. And this is no longer just a local problem. As we will see, easy urban access and rapid air travel can carry a virus from a remote community to another country or continent in just one leap. No wonder a tiny coronavirus can cause a global crisis!

Emerging human viruses may cause anything from a single infection, to a small outbreak, and on to an epidemic or pandemic. While an epidemic is defined as any unusual rise in the number of cases of an infection in a community, a pandemic is an epidemic that sweeps round the world, spreading on several continents at the same time. The main factors that determine whether an outbreak progresses to an epidemic and on to a pandemic are the virus's ability to infect and spread between humans, the availability of non-immune hosts within the virus's range, and the effectiveness of any precautionary measures taken to inhibit virus spread. This is measured by the R number, or case reproduction number. R is the average number of cases of an infection generated by one case (Figure 8). So if R is greater than 1

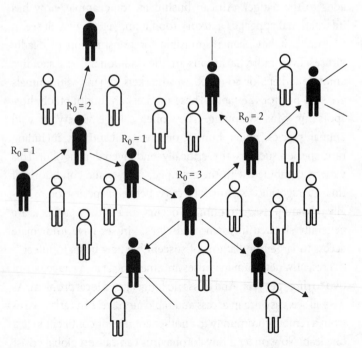

Figure 8. Ro: the basic reproduction number of an epidemic.

Table 2. Comparison of reproductive number (Ro) and herd immunity threshold (HIT, see page 94).

Disease	Transmission route	Ro	HIT %
Measles	Airborne	12–18	92–95
Rubella	Airborne	6–7	83–86
Smallpox	Airborne	5–7	80–86
Polio	Faecal–oral	5–7	80–86
Mumps	Airborne	4–7	75–86
Yellow fever	Vector	4.8	60–80
COVID-19	Airborne	2.5–4	60–75
SARS	Airborne	2–5	50–80
Ebola	Body fluids	1.5–2.5	33–60
Flu	Airborne	1.5–1.8	33–44

then the epidemic is growing, if R is less than 1 then it is waning, and if R is 1 then the epidemic is stable. The maximum number for R occurs at the beginning of an epidemic when everyone is susceptible and no interventions are in place. This value is called Ro, and is constant for any given virus or virus strain (Table 2).

In this chapter we look at groups of emerging viruses with very diverse R values, not forgetting that viruses may move up or down the scale as circumstances change.

1. Viruses that spread no further than a single individual, such as rabies and hantaviruses.
2. Viruses that cause sporadic epidemics after introduction to a human index case from their primary host. These viruses disappear from humans between epidemics. Examples include Ebola, Lassa fever, and the coronaviruses that cause severe acute respiratory syndrome (SARS-CoV) and Middle East respiratory syndrome (MERS-CoV).

3. Viruses that can subsequently circulate continuously among humans causing epidemics, such as yellow fever, Zika, and dengue viruses.

The pandemic viruses of the twenty-first century—HIV, flu, and SARS-CoV-2—are discussed in the next chapter.

Viruses unable to spread from person to person

Rabies virus

Rabies is an ancient affliction caused by a virus that is 100 per cent lethal (Figure 9). It survives in today's world despite relying on animals biting one another for its transmission between hosts (Figure 10). Rabies virus drives its victims into a frenzy that lasts several days, during which time the animal's saliva is loaded with virus ready to infect its next victim.

From the site of a bite by a rabid animal, rabies virus homes in on nerves in the skin and begins the long journey up the nerve fibre to the brain. While travelling it causes no symptoms at all—the wound heals and all seems well. But eventually, anything from seven days to several years later—a period determined by the distance from the original bite to the central nervous system—the virus reaches the brain. Here it causes inflammation (encephalitis) which induces bizarre behavioural changes. Wolves, for example, are usually solitary, shy animals, but once infected with rabies they may seek out the company of other wolves or even approach human habitation.

Humans suffering from rabies become hyperactive, delirious, and often violent. Severe muscle spasms, particularly induced by drinking any fluid, are the classic symptom of hydrophobia. This behaviour is interspersed with lucid intervals when recovery

A
DECLARATION
OF SVCH GREIVOVS
accidents as commonly follow
the biting of mad Dogges,
together with the cure
thereof,

BY
THOMAS SPACKMAN
Doctor *of* Physick.

LONDON
Printed for *Iohn Bill* 1613.

Figure 9. Title page from Sir Thomas Spackman's treatise on rabies of 1613.

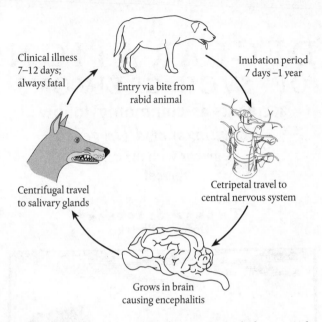

Clinical illness
7–12 days;
always fatal

Entry via bite from
rabid animal

Inubation period
7 days –1 year

Cetripetal travel to
central nervous system

Centrifugal travel
to salivary glands

Grows in brain
causing encephalitis

Figure 10. Rabies virus life cycle. Rabies virus travels from an infected bite along nerves to the brain where it grows and causes encephalitis. It then moves to organs including salivary glands where its presence in saliva allows its transmission to another host.

seems possible, but inevitably after a week or so sufferers sink into a coma and die. While all this is going on, the virus has been on the move again. From the brain it travels back along nerves, this time to tissues including the salivary glands, where it releases masses of new virus into the saliva.

In humans, rabies is a dead-end infection and so each new infection is introduced from an animal, commonly a rabid dog. But in the wild, the virus survives by circulating among foxes, wild and feral dogs, and bats, with the infectious cycle being perpetuated when a raving animal with a mouth full of rabies viruses bites an innocent victim. In fact, rabies is so common in certain African game reserves that it has completely devastated populations of wild dogs.

There is a vaccine against rabies, but although the virus has been eliminated from some island states, including the UK and Australia, global death rates from rabies are surprisingly high at around 70,000 annually, with the majority occurring in India.

Hantavirus

Like rabies virus, hantavirus does not spread between humans. A temporary fluctuation in climate was enough to cause the first known outbreak of hantavirus infection in the US, in 1993. This began when a previously healthy young couple were rushed to hospital in New Mexico with severe flu-like symptoms—fever, headache, cough, and difficulty in breathing. Both died rapidly and doctors were puzzled. Inquiries locally uncovered patients with similar symptoms seen in other hospitals in New Mexico and neighbouring Arizona and Colorado (Figure 11). Clearly there was a new and terrifying disease around, so the State Health Department and the National Centers for Disease Control and Prevention in Atlanta took over the investigation. By the end of 1995, a total of 119 people had suffered from the mystery disease and 58 had died.[1]

Scientists identified a new type of hantavirus which they named 'Sin nombre' (Spanish for 'without name'), and the disease it causes is called hantavirus pulmonary syndrome. They found no evidence of spread between humans, but a search among local wild animals identified the deer mouse as the virus carrier. About 20 per cent of these mice in New Mexico, Arizona, and Colorado carry the virus without developing disease. The virus is excreted in their saliva, faeces, and urine over

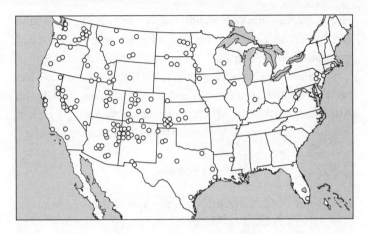

Figure 11. Map of the US showing cases of hantavirus infection until 1998.

long periods and passes easily from one generation of deer mice to the next.

The winter of 1992–3 was exceptionally mild, allowing more older mice than usual to survive. Then the following spring was particularly wet, producing an abundance of wild piñon nuts which deer mice like to eat. Consequently, they underwent a population explosion, spread more widely, and came into closer contact with humans, particularly in rural areas. Hantavirus spreads to humans by inhalation of infected material, that is, dust contaminated with mouse urine or faeces. Most people who suffered in the first outbreak either lived or worked in mouse-infested cabins, barns, or outhouses.

Since finding Sin nombre, scientists have identified the same virus in around 15 per cent of samples stored from people who died previously in the south-west US of unexplained flu-like illnesses, the earliest being in 1959. So the virus has clearly jumped species and infected humans before when conditions were favourable for a mouse population explosion.

Hantaviruses have since been found in other rodent species in the US, and worldwide, but, like Sin nombre, none is capable of spreading from one infected person to another. Presently, each person must be infected individually from the natural host—a factor that fortunately prevents large epidemics.

Viruses that cause sporadic outbreaks

Lassa fever virus

The first recorded victim of Lassa fever was an American nurse working at a mission station in the town of Lassa, north-east Nigeria, in 1969. She developed a severe flu-like illness and was eventually flown for treatment to the Evangel Hospital in Jos, where she died. The nurse who cared for her caught the virus and died 11 days later, and then the head nurse at the Evangel Hospital fell sick. She was flown to the US (in the first-class cabin of a commercial aircraft, with no special precautions taken!) where she slowly recovered. Virologists at Yale Arbovirus Unit, New Haven, US isolated a virus from her blood which they called Lassa fever virus. While working on this virus one of the scientists became ill and was only saved by an infusion of blood plasma from the nurse who had recently recovered. Five months later, a laboratory technician from the Yale Unit came down with Lassa fever and died—although he had never worked directly with the virus. This chain of events, coupled with several later, more devastating outbreaks in West Africa, gave Lassa its formidable reputation.

Lassa fever virus is endemic in West Africa, where it is carried by African brown rats. These rodents become infected in the nest at birth and excrete the virus freely in urine and faeces throughout their lives. They are common in and around towns and villages in sub-Saharan Africa, and during the rainy season they shelter in buildings, coming into closer contact with humans who become infected through touching contaminated objects or eating contaminated food. In these areas infection is relatively common, but spread between humans is unusual, requiring contact with body fluids from an infected person. This generally occurs in the hospital setting. In West Africa there are 300,000–500,000 cases annually, with around 5,000 deaths.

In 1989, the city of Chicago had a lucky escape. A man with flu-like symptoms at the height of a winter flu epidemic was, not surprisingly, assumed to have a severe case of flu. His case raised no suspicions for eight days, by which time he was seriously ill. Only then was he questioned closely and his story emerged. He had attended his mother's funeral in Nigeria two weeks previously, and ten days after that his father died. It emerged that both his parents had illnesses similar to his own, and this set alarm bells ringing. He was suffering from Lassa fever and he died shortly after the diagnosis was made. During his illness, when he was presumably excreting the virus, he had been in contact with 102 people, all of whom were traced, but fortunately none developed the disease.

Ebola disease virus (EDV)

The first recorded Ebola outbreak occurred in Yambuku, a remote village in North Zaire (now the Democratic Republic of the Congo or DRC), in 1976. In August of that year the village school teacher, who had just returned from a trip into the bush, arrived at the local Catholic mission hospital complaining of headache and fever. The Belgian nuns who ran the mission hospital gave him an injection of antimalaria drugs, but he did not have malaria; he had Ebola haemorrhagic fever. Within a few days he was dead, but not before infecting several others with the deadly virus, including his family and the nuns who nursed him. This started an epidemic which lasted three months, affected 318 people, and killed 280 of them.

The disease, after a few days of non-specific, flu-like symptoms develops into a haemorrhagic phase with vomiting, diarrhoea, a skin rash, and, in severe cases, the characteristic bleeding, with blood oozing from the ears, eyes, and gums, and into vital internal organs. The most common cause of death is probably dehydration due to fluid loss from diarrhoea and vomiting, but massive haemorrhage into the gastrointestinal tract can also prove fatal. On witnessing this dreadful new disease take hold, the panicking nuns called for backup and a team arrived from Kinshasa, capital of Zaire, to help out. They took blood samples from victims and sent them to colleagues at the Prince Leopold Institute of Tropical Medicine in Belgium. Ebola disease virus (EDV) was isolated from these blood samples, prompting an International Commission of experts to travel to Yambuku to investigate. The Commission arrived in October 1976 and immediately set about house-to-house searches, following chains of infection to unravel the epidemiology of the disease. They identified the school teacher as the index case and the mission hospital as the epicentre of the outbreak, and decided that the most likely transmission route of the virus was contact with a victim's bodily fluids, be it blood, saliva, vomit, faeces, or urine. The incubation period between infection and disease was six to nine days and the death rate around 88 per cent.

Major risk factors for infection included attending the mission hospital (where injections were administered with unsterile equipment) and participating in the traditional funeral rites of an Ebola victim, which included touching and washing the body.[2]

Ebola's sudden appearance in a remote area, its rapid spread from the index case, and its disappearance once most local people had been infected are typical of this virus. Over the next 40 years around 20 such outbreaks were recorded, always in either the DRC, Uganda, Gabon, or Sudan (Figure 12). In each epidemic, the virus has infected an index case from an unknown (presumed animal) reservoir in the rainforest.

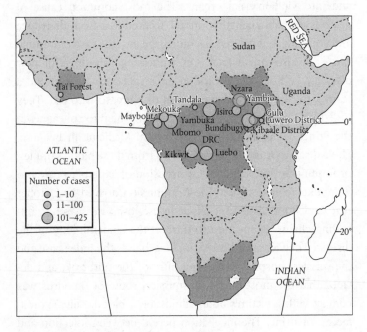

Figure 12. Map of Ebola outbreaks in Africa.

With rapid case identification and isolation, strict barrier nursing, contact tracing, and quarantining for 21 days, the outbreaks were successfully curtailed and the virus eliminated. But despite exhaustive trapping and testing of wildlife in the areas involved, the primary host for Ebola virus was not identified.

Ro for Ebola is 1.5; this relatively low figure is due to the rapidly developing, debilitating symptoms that prevent sufferers from moving far from home. This, combined with the frighteningly high mortality rate (50–90 per cent), led to the belief that the virus would not cause widespread epidemics or become endemic in humans. It typically caused explosive but localized epidemics in isolated areas that died out when the virus had exhausted its supply of new hosts. And this was indeed the case until March 2014, when an outbreak began in a remote corner of Guinea, West Africa, close to its borders with Sierra Leone and Liberia.

Because Ebola had never occurred in Guinea before, it was not immediately recognized, and the virus spread to neighbouring Sierra Leone and Liberia before the alarm was raised. Even then many factors conspired to allow the virus to get the upper hand. These three countries are among the poorest in the world, and each had just emerged from long periods of unrest. Health services were unable to cope with the influx of cases, and because personal protective equipment (PPE) was in very short supply, many healthcare workers caught and died of the disease. Furthermore, people had a deep mistrust of their governments and of western medicine, so instead of following the advice being given to them, they hid Ebola cases in the villages, turned to traditional healers for medicines, and performed traditional burial rites including the custom of touching and washing the highly infectious corpse.

One particular example of the problem occurred in Aberdeen, a coastal slum area of Freetown, capital of Sierra Leone. Here, makeshift shacks on or near the beach were home to fishermen and their families, with around 10,000 living without clean water or waste disposal and only four toilets to serve the whole community. On going out to sea, the small fishing boats from Aberdeen often met up with similar vessels from further up the coast. This contact first brought Ebola virus to Aberdeen, with more than 20 cases occurring simultaneously. That might have been the end of the affair, but one fisherman, who was quarantined when the first cases appeared, decided to head home to his village some 200 kilometres away when he began to feel unwell. He escaped quarantine and passed through at least 12 checkpoints undetected by hiding in the back of a truck. On reaching home he was cared for by his family until he died of the disease. Large numbers gathered for his funeral and participated in ritual washing of the body. They then poured the washing water over themselves and the children played in the puddles. Twenty-four people, both children and adults, caught Ebola at this event.

It is no wonder that for six months the virus spread uncontrollably, reaching all three capital cities as well as Nigeria, Mali, and Senegal. It was not until August 2014 that the WHO announced a Public Health Emergency of International Concern in West Africa, and set about coordinating the response. Key to the success of this intervention rather late in the day were the recruitment of international volunteer healthcare workers, the rapid construction of treatment centres and diagnostic laboratories, ample provision of PPE, and public engagement to explain the nature of the disease and the necessary precautions required to eliminate it. With no drugs or vaccines available to prevent or treat Ebola, the traditional methods of track, trace, and isolate were employed with beneficial effect. In fact, there *were* a few vaccines in the pipeline, but these had been shelved at an early stage of production through lack of funds, and also to await an Ebola outbreak for efficacy trials. These trials finally got underway in mid 2015 when the peak of the epidemic was already passed, but so-called ring vaccination of all contacts of cases did demonstrate that the vaccines could prevent infection. This international effort, with essential support from local community workers, eventually controlled virus spread.

In January 2016 the epidemic was declared over, but not before infected volunteer healthcare workers carried the virus to their home countries, with person-to-person spread reported in the US and Spain—the first outside Africa. Overall, there had been 28,637 cases reported, with a 40 per cent death rate (Figure 13).

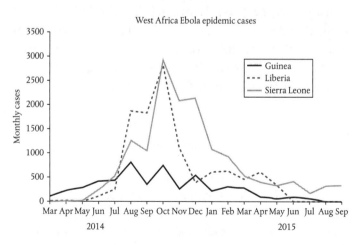

Figure 13. Ebola cases in the 2014–16 epidemic.

EDV had once again been eliminated from humans, but it was a close-run contest. For although the outbreak began in a rural setting, with the extensive road networks that exist in all three countries connecting villages with towns and cities, the virus rapidly entered their capital cities and became an urban dweller. Here it spread easily among the crowds, so becoming much more difficult to control. And for the first time it took to the air, just an indication of what the future might hold. Then, just as the epidemic was finally drawing to a close, some worrying cases came to light. Several months after the initial infection, virus was detected in two patients suffering long-term effects of Ebola. One had virus in the eye,[3] the other in the brain.[4] In addition, virus was detected in the semen of a recovered case and this had been transmitted to a partner, causing Ebola disease.[5] These may be rare, anecdotal accounts, but if substantiated, they suggest that the virus could establish a reservoir in humans, and that would certainly be a game changer for Ebola.

Again intensive searches have failed to identify Ebola virus's primary host. The virus is known to infect apes, and a variety of domestic animals, which can pass the virus to humans; but as these all suffer from the disease, they are not the animal reservoir. The most likely culprit is a bat species, particularly fruit bats, but so far the virus has not been isolated from any bat species.

Coronaviruses

The first coronavirus was isolated in the 1930s from chickens suffering from a respiratory infection, but this family of viruses was not so named until 30 years later, when in 1960 the first human coronavirus was identified. This virus is one cause of the common cold, and its discovery is discussed in Chapter 5.

More than 40 years then passed before another, much more aggressive coronavirus infected humans. And within the following 20 years two more human coronaviruses had emerged. Unlike the common cold-causing human coronaviruses, all three new additions to the family cause severe acute respiratory disease with high fatality rates. So where did these viruses come from? How and why did they infect humans?

Severe Acute Respiratory Syndrome Coronavirus (SARS-CoV), causing the disease 'SARS', emerged in Foshan, a city of 7 million inhabitants in Guangdong Province, China, in November 2002. It caused an outbreak of atypical pneumonia that spread locally around Foshan, before reaching Guangzhou, capital city of Guangdong Province, in January 2003. Just one month later, a doctor who had treated SARS patients in Guangdong Province travelled to Hong Kong to attend a wedding. He stayed one night in a hotel in Hong Kong before he was taken ill and admitted to hospital. He died of SARS a few days later, but by then he had sparked an epidemic in Hong Kong and initiated the virus's global spread. The doctor infected at least 17 other hotel guests who then boarded international flights and carried the virus to several other countries, sparking epidemics in Canada, Vietnam, and Singapore (Figure 14).

In Hong Kong the doctor's admission to the Prince of Wales Hospital initiated an outbreak among patients, doctors, nurses, students, and visitors, eventually infecting around 100 people. One of these carried the virus to the Amory Gardens private housing estate while visiting his brother, so seeding an outbreak involving more than 300 cases and 42 deaths. This outbreak was only brought under control after all residents were evacuated to an isolation facility while the building was sterilized. The investigation that followed blamed rapid spread of the virus through the block by a rather unusual route. Although SARS-CoV usually

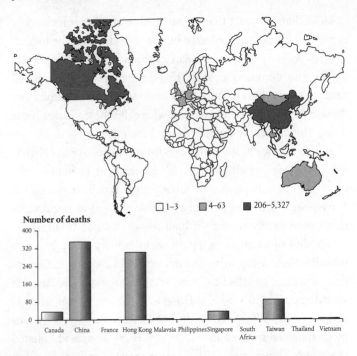

Figure 14. Global map of SARS-CoV spread during the epidemic in 2002–3, the shading of the map indicates the number of cases reported.

spreads through the air, diarrhoea is a feature of the acute disease and the virus is present in faeces of sufferers. This, combined with a partially blocked sewage system and over-powerful exhaust fans in the bathrooms, created the potential for airborne virus from toilets to enter air shafts and thereby sneak into apartments throughout the building.[6]

SARS-CoV particles in inhaled airborne mucus droplets infect cells lining the respiratory tract. After an incubation period of between two and seven days, the infection causes fever, malaise, muscle aches, a cough, and sometimes diarrhoea. This progresses to viral pneumonia in around 20 per cent of cases, often requiring intensive care and mechanical ventilation. Deaths were most common among the elderly, with an estimated overall rate of around 10 per cent.

Ro for SARS-CoV is 2–4, so the wave of international dissemination from Hong Kong could easily have progressed to a pandemic, particularly if the virus had been carried to middle- or low-income countries. As it was, this scenario was prevented by swift and decisive action by the WHO. Like the earlier Ebola outbreaks, no vaccines were available to prevent virus infection, so it was the traditional methods that saved the day—rapid case identification and isolation, along with efficient tracking, tracing, and isolating all contacts, use of PPE, as well as the evacuation of a housing estate.

Fortunately, these efforts were augmented by several features of the disease itself which, once appreciated, aided its control. First, virtually everyone infected with SARS-CoV becomes unwell, and second, those infected are not infectious before they manifest symptoms. In practice, this means that there are no asymptomatic virus carriers who could unwittingly spread the virus. Third, during the disease, the virus is mainly produced deep in the lungs and shed into the air in mucus droplets by coughing These droplets are relatively heavy and so do not spread far from the patient. Thus, those most at risk of infection are close contacts such as medical carers and family members. So, when all those showing clinical signs of the disease and their contacts were swiftly identified and isolated, and those in close contact with SARS cases used PPE, the epidemics were curtailed, and by July 2003 SARS was gone. During the epidemic the virus reached 26 countries, causing more than 8,000 cases, with a final death toll of around 800.

As several of the first SARS cases to be identified were people who worked in a live animal market in Foshan, this was where the search for the primary animal host for the virus was concentrated. Here such animal species as the Himalayan palm civet cat and the racoon dog tested positive for the virus, but the natural reservoir of the virus was later identified as the Chinese horseshoe bat. The assumption is that, on occasions, virus spillover occurs from carrier bats to other animals in the market, where they are confined in overcrowded cages. Market traders handling these animals are then in danger of acquiring the virus from these intermediate hosts. Retrospective testing of market traders supported this assumption by demonstrating antibodies against SARS-CoV in 40 per cent of wild animal traders and 25 per cent of slaughterers, compared to just 5 per cent of vegetable traders.[7]

Since July 2003, SARS-CoV has reappeared on three occasions, involving virology laboratories in Taiwan, Singapore, and China. All were caused by inadequate or breached safety procedures in laboratories known to handle the virus. In two of these outbreaks the virus did not spread beyond the index case. But an outbreak originating in China's National Institute of Virology Laboratory in Beijing resulted in nine cases and one death. The outbreak began with two cases in graduate students from the Institute, one of whom later travelled from Beijing to her home in Anhui Province in eastern China while feeling unwell. The virus spread to her mother, who died, and the nurse of the graduate student, and then from the nurse to five further contacts.[8]

Laboratory outbreaks of SARS can surely be contained, but with wet markets still commonplace in the Far East, reemergence of this virus is a real possibility.

The disease called Middle East Respiratory Syndrome (MERS), caused by MERS-coronavirus (MERS-CoV), was first recognized in 2012 in Jeddah, Saudi Arabia, where a small outbreak of a highly lethal SARS-like illness occurred. Symptoms include fever, cough, shortness of breath, and sometimes diarrhoea. Progression to pneumonia is common, particularly in the elderly and those with underlying health problems. Sufferers often require hospitalization and mechanical ventilation. The overall death rate is approximately 35 per cent.

Since its emergence there have been several small outbreaks of MERS, with approximately 2,500 reported cases in all. Around 80 per cent of these cases occurred in Saudi Arabia, and all those diagnosed outside the Arabian Peninsula have been in travellers infected within that area.

The coronavirus that causes MERS, MERS-CoV, was identified during the first outbreak and proved to be similar to, but distinct from, SARS-CoV. It is thought that this virus originated in bats, but human infection is acquired from dromedary camels. It seems that this intermediate host acquired the virus from bats some time ago and it now circulates continuously among these camels, carried as an asymptomatic infection in the nasal passages. Spillover to humans most often occurs through direct or indirect contact with camels, hence the alternative name for MERS, 'camel flu'.

Spread of MERS between humans requires close contact and is uncommon outside the hospital setting. R0 for MERS is less than one (estimates vary from 0.4 to 0.9), meaning that outbreaks are generally small and, once recognized, are relatively easy to control given the appropriate use of protective equipment. Nevertheless, two fairly large outbreaks have occurred—one in Jeddah and Riyadh, Saudi Arabia, in 2014, involving 180 cases and lasting around 70 days; the other in South Korea in 2015, with 186 cases over 55 days. Both outbreaks were hospital-based, and in South Korea virus spread was fuelled by a delay in implementation of effective countermeasures on the assumption that if Ro was below one the virus would not spread.[9]

No vaccine is presently available to prevent MERS, and although the WHO declares that the virus is unlikely to pose a global threat, it appears on the WHO's list of priority diseases for research and development.

Emerging viruses that circulate continuously among humans

The viruses described in this section are all arboviruses, meaning that they are spread by insects, in these cases, mosquitoes. These viruses all originated in primates, with spillover to humans occurring many years ago. However, as we will see, some arboviruses, like yellow fever virus, have retained their connection with the primary hosts. Although the arboviruses described here are not new to humans, they are classed as emerging, or, more accurately, re-emerging, viruses, because they are presently increasing their global range.

An arbovirus's geographical range is determined by the availability of its human host as well as its vector, and over the past century both these parameters have changed. First, the human population explosion ensures these viruses have an almost unlimited supply of susceptible hosts, while rapid, long-distance travel disseminates virus-carrying hosts to new territories.

Second, *Aedes* mosquitoes depend on very specific constraints for successful completion of their reproductive cycle. High ambient temperature is required for females to incubate their eggs successfully, while high rainfall is essential for the aquatic larval stage. This has previously restricted the insects to tropical regions, but climate change has extended their range substantially and the viruses that depend on them have followed along. Values of Ro for arboviruses are somewhat variable because of the viruses' dependence on a vector. In their traditional tropical homelands Ro averages at 4.2 for yellow fever virus, 4.7 for dengue fever virus, and 3.0 for Zika virus. However, these values are lower when the viruses spread in temperate zones because their vectors are less abundant.

Yellow fever virus

Yellow fever was the first human disease demonstrated to be caused by a virus, but it had been responsible for human epidemics for several hundred years before their cause was discovered. Yellow fever was first reported in Havana, Cuba in 1648, and it soon took hold in South and North America. The virus probably arrived in the Americas from its roots in Africa aboard slave ships bound for Barbados in the mid-seventeenth century. Virus-carrying mosquitoes survived on board by breeding in the ships' water barrels and continued to spread the virus among passengers and crew during the long voyage. Once in port, the mosquitoes moved inland and established an outpost in local rainforests.

Thereafter, yellow fever epidemics occurred regularly in riverside towns with a warm, wet climate like Memphis and New Orleans, and also in ports like New York and Philadelphia. A devastating outbreak in Philadelphia in 1793 killed over

4,000 people—a tenth of the population—in just four months. Victims developed a high fever, severe head and joint aches, bleeding, nausea, and bloody vomiting. Many died at this stage from internal bleeding but others went on to develop the jaundice which gives the disease its name, often dying of liver, heart, or kidney failure.

Of course, at the time, the way yellow fever spread between its victims was a mystery. It did not seem to pass directly from one person to another, or to contaminate food or drinking water, but it could be carried from one community to another by infected people fleeing a disease-ridden town. Its appearance in ports often coincided with the arrival of ships, yet placing these in quarantine did not prevent the disease spreading inland.

In 1898, during the Spanish–American war, an epidemic of yellow fever struck the US forces occupying Cuba, with 231 soldiers dying as a result. This catastrophe finally persuaded the American government to take action against this much-feared disease, which had by then plagued the Americas for over two centuries.

In 1900, the army set up the US Yellow Fever Commission and a team led by Walter Reed, Professor of Bacteriology at the US Army Medical School, headed for Havana to investigate. They used human volunteers to determine whether mosquitoes spread the disease, and three of the group offered to act as human guinea pigs. They were eventually convinced that mosquitoes were the vector, but not before two of the party—James Carroll (originally Reed's laboratory assistant) and Jesse Lazear (head of the clinical laboratories at Johns Hopkins Medical School)—contracted the disease. Carroll recovered, while Lazear died.

The idea that mosquitoes could spread a disease was entirely new in 1900, and at first it was ridiculed. However, after further carefully controlled experiments with more human volunteers, people were convinced, and preventive measures were organized. In 1901, a team from the US Army under Major William Gorgas set about ridding Havana of mosquitoes by removing their breeding grounds. They either drained all standing water or covered its surface with oil, and this successfully controlled yellow fever in the city. Yellow fever virus was finally isolated in 1927.

Monkeys of the African and South American rainforests act as reservoirs of yellow fever virus, spread among them by mosquito species of the genera *Aedes* or *Haemagogus*. This stage is called the sylvatic (or jungle) cycle. Once she has taken her blood meal from an infected monkey, a female mosquito injects the virus into her next victim, whether a monkey or a human—perhaps a logger or tourist in the rainforest. But the virus is not just a passive traveller in the mosquito; it multiplies in her intestine then travels to the salivary glands, where it multiplies again before being injected into the next victim. The virus causes no inconvenience to the mosquito that might jeopardize its chances of successful transmission. Once mosquitoes have transmitted yellow fever from monkey to human they can then transfer it directly between humans, initiating the urban cycle. This cycle amplifies effortlessly in crowded, urban situations where day-flying mosquitoes find an abundance of susceptible hosts, so causing an epidemic (Figure 15). A vaccine against yellow fever virus was developed in 1937 and has proved very effective in preventing the disease. This vaccine is still the mainstay of yellow fever control.

Today, the WHO records endemic yellow fever in 47 countries in tropical areas of Africa and South America, causing around 200,000 cases of severe disease and 30,000 deaths annually. The virus is presently reemerging, mainly due to a

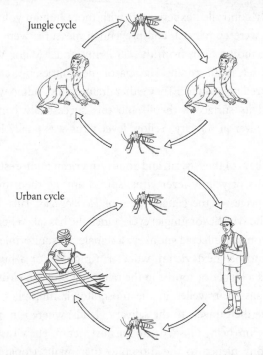

Jungle cycle

Urban cycle

Figure 15. Yellow fever transmission cycle. In the rainforest the virus is carried from monkey to monkey by a mosquito. If a virus-carrying mosquito bites and infects a human, then there is the potential for spread among humans via the mosquito vector.

waning of vaccine coverage. Furthermore, the mosquito vector is benefitting from increasing urban populations in endemic areas which provide the overcrowded living conditions in which it thrives. As long as this continues yellow fever will be a global problem, with the real threat of imported cases sparking epidemics in other countries and continents. As a result of two large epidemics of yellow fever in the Democratic Republic of the Congo in 2016 which exported cases to Angola, Kenya, and China, the WHO launched the Eliminate Yellow Fever Epidemics

(EYE) Strategy in 2017, with the declared aims of protecting at-risk populations, containing outbreaks rapidly, and preventing international spread. The combined linchpins of the strategy are vaccination and vector control, which together are affordable and extremely effective. However, with its natural reservoir in several species of monkey in Africa and South America, the virus cannot be entirely eliminated.

Dengue fever virus

Dengue fever virus, once restricted to Asia, is now rife in Africa, South and Central America, the Pacific Islands, and Northern Australia. The virus is presently endemic in more than 100 countries with an estimated 3.9 billion people at risk of infection, 70 per cent of whom live in Asia. Imported cases reached Europe for the first time in 2010 with onward transmission of the virus causing localized outbreaks.

Although dengue fever is often subclinical, it also causes a range of symptoms—from a flu-like illness with severe headache, muscle, bone, and joint pains dubbed 'break-bone fever' to a potentially lethal haemorrhagic fever called 'severe dengue'. There is no specific treatment for dengue, but with modern medical support the fatality rate is around 1 per cent.

Like yellow fever, dengue virus is primarily transmitted by *Aedes* mosquitoes with an amplification cycle within the insect's salivary gland. And this insect thrives in urban and suburban areas, breeding rapidly in overcrowded, chaotic urban developments where ponds, puddles, drains, drinking water reservoirs, and air conditioners form ideal incubators for developing larvae. Inside the rims of old rubber tyres is a particularly favoured breeding ground, and there are plenty to be found in large cities.

There is no vaccine against dengue suitable for widespread use, so its control depends on controlling the vector with insecticide sprays.

Zika virus

Zika virus rose from obscurity to hit the headlines in 2015–16 when it caused large epidemics in South America and the Caribbean islands. Like the viruses causing yellow fever and dengue, Zika virus is not new to humans. Zika was first isolated in 1947 from rhesus monkeys in the Zika forest in Uganda. At the time it was restricted to Africa and Asia and was known to be spread between monkeys by *Aedes* mosquitoes, with the occasional transfer to humans via a bite from a virus-carrying mosquito. But this was of no particular concern because infection was either asymptomatic or caused mild flu-like symptoms with no known complications.

In 2007 Zika virus began to expand its territory. Leaving behind the monkeys of the Zika forest, the virus island-hopped across the Pacific from Asia to South America inside infected travellers (Figure 16). Arriving in Brazil in 2015, Zika virus caused a large epidemic during which a rise in birth defects was noted. In particular, microcephaly (meaning a small head with an underdeveloped brain) was being reported at higher than usual rates in babies born to women infected during pregnancy. In addition, occasional cases of neurological damage (Guillain–Barré syndrome) were noticed in infected adults. In February 2016, when these severe sequelae were conclusively shown to be caused directly by the virus, the WHO declared a Public Health Emergency of International Concern. At the time one particular worry was the Olympic games, due to be held in Rio de Janeiro, Brazil in summer 2016. This would bring millions of people to

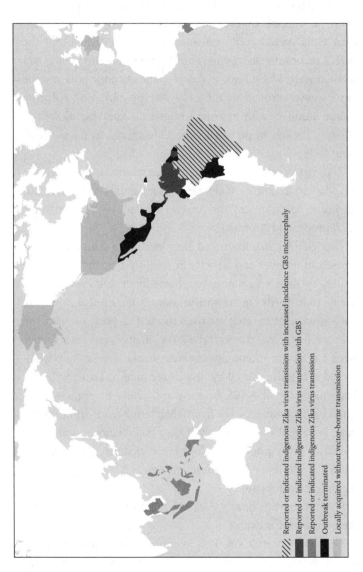

//// Reported or indicated indigenous Zika virus transmission with increased incidence GBS microcephaly

▨ Reported or indicated indigenous Zika virus transmission with GBS

▦ Reported or indicated indigenous Zika virus transmission

■ Outbreak terminated

░ Locally acquired without vector-borne transmission

Figure 16. Global map of Zika virus spread.

Rio, athletes and spectators alike, all vulnerable to the virus. With no vaccine or drugs to prevent or treat the infection, Zika virus spread could not (and cannot) be halted, but the WHO recommended going ahead with the games. Pregnant women were advised not to travel to the endemic area and the virus was controlled in the only way possible—by reducing vector numbers with industrial insecticide spraying as well as advising the use of personal insecticide sprays. In the event all went well.

Zika virus continues to spread more widely, and now that it has adapted to utilize a wider range of mosquito vectors, including non-tropical species, the virus is expected to spread on to the US, Europe, and Australasia.

This chapter has illustrated how emerging viruses have increased in number and ferocity in recent years, and how these viruses by their very nature unashamedly exploit every opportunity that opens up to spread among us, be it a cramped, unhygienic wet market, an overcrowded airport, or with the help of discarded rubber tyres! For many years the danger signs have been mounting, apparent in the increasingly widespread epidemics caused by new or reemerging human viruses. The epidemics of SARS in 2003 and Ebola in 2014–16 both spread on continents far from their emergence. Both could have gone global, particularly if they had chanced to arrive in countries with poorly developed surveillance networks and medical facilities. What's more, arboviruses are rampant and seemingly unstoppable.

So what lessons have we learnt? What can we do to control these viruses effectively?

The first is prevention. This requires strengthening of global health security by bolstering global health infrastructure, monitoring for emerging viruses, control of flashpoints, and research into, and preparation of, vaccines against known emerging viruses.

The second is the control of outbreaks. This requires rapid, decisive global action by experienced international response teams with local involvement and stockpiling of PPE.

Unfortunately, to date the story is one of missed opportunities, all leading to the emergence of a particularly effective killer virus—SARS-CoV-2—and the COVID-19 pandemic. If only we had heeded the warnings we would undoubtedly have been better prepared to combat this dreadful threat.

As Benjamin Franklin said, 'By failing to prepare, you are preparing to fail'.

4

TWENTY-FIRST-CENTURY PANDEMICS

Before we are even a quarter of the way through the twenty-first century, we have experienced three pandemics, with millions of lives lost. The first of these is HIV-1, which began in the twentieth century and is still ongoing, while H1N1/09 swine flu and COVID-19 arose in 2009 and 2019, respectively. All three pandemics are caused by zoonotic RNA viruses that appeared without warning and had spread uncontrollably before the alarm was raised. While all three are entirely new to humans and are therefore classed as emerging infections, other strains of the flu virus have infected us for many centuries. In this chapter we look at where and how these viruses jumped to humans, and how and why they spread around the world.

Human immunodeficiency virus (HIV)

The story of the emergence and pandemic spread of HIV is an extraordinary one. Only once before has a microbe primarily transmitted by sexual contact caused a pandemic, and that was syphilis, caused by the bacterium *Treponema pallidum*. Syphilis hit Europe like a thunderbolt at the end of the fifteenth century, but, unlike HIV, it was immediately obvious because it caused an acute—and, at the time, deadly—disease.

Yet when the first cases of disease caused by HIV were recognized, in 1981, the causative virus had been silently spreading in humans for more than 40 years. The first inkling came when the Centers for Disease Control in Atlanta, US reported that 'five young men, all active homosexuals, were treated for biopsy-confirmed *Pneumocystis carinii* (now renamed *P. jirovecii*) pneumonia in three different hospitals in Los Angeles, California'.[1] Then, just one month later, came the description of 26 cases of Kaposi's sarcoma—until then an extremely rare tumour in the US—again all in previously healthy, gay men.[2]

The appearance of *Pneumocystis* and Kaposi's sarcoma, both generally only seen in people with severely suppressed immunity, in previously fit and healthy, young gay men, had doctors baffled and prompted a countrywide survey. This indicated that they were looking at an entirely new disease in the US that, as the reports suggested, manifested as a severe immunosuppression and was mainly restricted to gay men. The disease presented clinically as a combination of life-threatening opportunistic infections, often caused by microbes that are harmless to those with a competent immune system. It was named acquired immune deficiency syndrome (AIDS), and despite treatment for the multiple infections, as there was no cure for the immune deficiency the disease was 100 per cent fatal. Early on, the AIDS outbreak was centred on California and New York, but very soon cases were popping up in Haiti, Europe, Australia, and many other locations. It had become a pandemic, and the race was on to find the cause.

While gay communities suffered the devastating loss of hundreds of previously healthy young men, finger-pointing and discrimination were rife. There was much debate about the cause of AIDS, some arguing that it was the effect of drugs such as amyl and butyl nitrite in 'poppers', used as inhalant aphrodisiacs. And although a viral cause was always mooted, this became much more likely when AIDS cases were diagnosed in groups other than gay men, including intravenous drug users, blood transfusion recipients, and haemophiliacs. It took two years of hard slog, but in 1983 human immunodeficiency virus (HIV), a retrovirus, was finally isolated from the lymph gland of a French gay man.[3] But several more years passed before there was enough evidence to convince everyone that HIV caused AIDS.

HIV transmission

HIV is primarily spread by sexual intercourse, be it heterosexual or homosexual contact. Both male and female genital secretions from infected individuals contain the virus, and injury to the lining of the genital tract or ulceration caused by other sexually transmitted diseases increases the risk of infection. In contrast, practices which prevent injury, such as circumcision (possibly because of thickening of the skin on the tip of the penis) and condom use, protect from infection.

HIV infects CD4 T cells that circulate in the blood and so the virus can spread directly to uninfected people via a blood transfusion. This transmission route allows the virus direct access to circulating CD4 T cells, but with millions of viruses in every millilitre of blood, any contamination, be it from an infected needle or during the birth of a baby of an infected mother, may be enough to initiate a new infection.

One of the tragedies of the HIV pandemic was infection of haemophiliacs through contaminated human clotting factor VIII, required by them to prevent bleeding. During the 1980s more than 1,000 haemophiliacs in the UK, and many more worldwide, contracted HIV by this route, and the death rate among them rose tenfold. However, as soon as the problem was recognized it was relatively easy to control, by restriction of donors and by HIV testing of blood and blood products; this soon became routine in almost every country.

Also at risk were healthcare workers, in whom accidental exposure could occur through a needle-stick injury. One such was Malon Johnson, a pathologist at Vanderbilt University, Tennesssee, US, who specialized in brain diseases. At 8.00 p.m. one evening in September 1992, his phone rang. An AIDS patient had just died and Johnson decided to carry out the post-mortem examination straight away. As always, he took all the right precautions—protective clothing, face mask and shield, arm protectors, and double latex gloves—but on this occasion it wasn't enough. Surrounded by his kit of lethal weapons—saws, knives, scissors, forceps—he set to work to remove the dead man's brain. While peeling away the skin his bloody scalpel slipped and cut deep into his thumb. Again he followed all the prescribed guidelines—a thorough wash, encouraging bleeding, and applying disinfectant. The next day he reported to the clinic and signed up for regular blood tests, but that was it: he was HIV positive. In his book, called *Working on a Miracle*, Johnson describes his experience in graphic detail, ending in 1996 when he is healthy and taking antiretroviral drugs.[4]

Intravenous drug use is another route for HIV infection, and this is difficult to control, mainly because the practice is illegal. In the early 1980s, drug users were not supplied with sterile needles and syringes so old injecting equipment was shared and reused without a thought for its sterility. As a result, HIV, along with other blood-borne viruses, passed freely from the blood of one needle-user to another.

HIV emergence

Like most emerging viruses, HIV is zoonotic, crossing the species barrier from animal to humans, but the particular animal, or as it turned out, animals, involved proved difficult to track down. All along there were plenty of scare stories—was it a man-made virus released by the US Government? A contaminant of polio vaccine? Or perhaps a virus from an as yet undiscovered animal lurking in the African jungle?

Because AIDS was first recognized in the US, it took some time for the link with Africa to be uncovered. But as increasing numbers of Africans residing in Europe were diagnosed with the disease, the light eventually dawned. In 1984, a group of infectious disease experts undertook a fact-finding mission to Kinshasa, capital of the DRC, and Kigali, capital of Rwanda. In both cities they found hospital wards full of cases of so-called slim disease, a severe chronic wasting disease with malaise, fever, and loss of appetite—clearly AIDS by another name. So HIV had already reached epidemic proportions in these cities, where it affected men and women equally; and at last came the recognition that HIV usually spread by heterosexual contact.

Surprisingly, there turned out to be two types of HIV: HIV-1 and -2, as well as several HIV-1 subtypes. The epicentres for both epidemics were in Africa, and by searching through banks of stored blood samples, scientists found evidence of HIV-1 infection in humans in Central Africa dating back to the 1950s, and HIV-2 in West Africa since the 1960s. At that time, levels of both infections were too low to cause epidemics, but since then, whereas HIV-2 remained localized to West Africa, HIV-1 has spread globally.

So the search for the primary hosts of HIV-1 and HIV-2 moved to Africa. Many African primate species carry viruses known as simian immunodeficiency viruses, or SIVs, which belong to the same retrovirus family as the HIVs, and scientists soon realized that HIV-2 was so closely related to the SIV of the sooty mangabey monkey that this was its likely origin. This primate is sometimes hunted for food or kept as a pet in West Africa, so over the years humans probably occasionally picked up the virus though infected bites or cuts. These few scattered cases would have attracted no particular attention, but eventually the virus established a sufficient toe-hold to spread successfully between humans, causing an epidemic.

In contrast, chimpanzees eventually proved to be the source of HIV-1, and of the four chimp subspecies, it is the SIV from the black-faced central chimp, subspecies *Pan troglodytes troglodytes*, that is the closest relative of HIV-1 and therefore its direct ancestor.

The genetic divergence between HIV-1 and this chimp SIV suggests that the chimp virus jumped to humans on more than one occasion back in the 1930s, probably during hunting, slaughtering, and dismembering the animals for bushmeat. The most likely location for this transfer has been identified as the south-east corner of Cameroon, where members of one particular tribe of chimps carry the SIV strain that is most closely related to HIV-1. Yet the virus did not spread outside this remote area until much later.

HIV-1's transcontinental journey

Urbanization and human migration played a key role in the spread of HIV-1 in Africa. Increasingly people moved from rural to urban areas in search of work, and HIV moved with this migrant population to become a city dweller. The earliest HIV-1 positive stored blood sample found in Africa is from 1959 in Kinshasa, revealing this city as the epicentre for HIV-1 dissemination.

In the 1950s there were few roads in rural Africa, and so rivers were the highways along which people, domestic animals, and merchandise moved. It is assumed that HIV-1 arising in southeast Cameroon travelled along the local Sanga river, a tributary of the Congo river, which then took it all the way to Kinshasa, doing so inside one person, or perhaps a series of infected people. Once there the city offered the virus great potential for spread between individuals. Those with multiple sexual partners, such as commercial sex workers, became infected early on and acted as a focus for dissemination of the virus. The infected population grew rapidly and HIV-1 became epidemic in the region by the late 1970s. By 1986, an estimated 1–2 million Africans were HIV-1 positive, including many young children infected at birth from their virus-carrying mothers.

In the early 1980s, a report was published describing a cluster of AIDS cases among Haitians living in Miami, US, and this prompted investigations in Haiti, situated on the Caribbean island of Hispaniola. These revealed that the first AIDS cases (uncovered retrospectively) in both the US and Haiti appeared at around the same time—between 1978 and 1979. This information generated a flurry of studies looking at the genetic diversity of early viruses isolated from AIDS sufferers from the US, South America, the Caribbean, Europe, Africa, and Asia. The results of these studies uncovered HIV-1's amazing journey from Africa to the US, heralding its global spread.

From its base in Kinshasa, HIV-1 took one enormous leap to Haiti inside a single infected person. No one knows who the virus-carrier was, but at the time many Haitian United Nations

Aid Workers were recruited to the DRC because of their shared language—French. So it is likely that one of these workers, returning home to Haiti, unknowingly carried the virus with them.

Arriving in Haiti in 1966, HIV-1 sparked a silent epidemic on the island and was at some point picked up by an American gay man. Again no one knows who this was, but as the island state was a developing gay holiday destination at the time, it was probably one of these holidaymakers. From here he transported the virus to the US around 1969, where it spread among the gay population, remaining hidden for over a decade before AIDS was recognized. Meanwhile, this focus of infection was HIV-1's launch pad for Europe and the rest of the world.

So, silently, HIV-1 infection reached frighteningly high levels before the problem was fully appreciated. In the space of 40 years it had mushroomed from being an unrecognized, isolated, rural infection of Central Africa into a pandemic that has now infected 75.7 million people worldwide, with 32.7 million deaths (Figures 17 and 18).

We will consider persistent HIV-1 infection and treatment and prevention strategies in Chapters 6 and 8, respectively.

Influenza (flu) virus

In January 2009 the world was hit by a pandemic of swine flu, so called because its genetic makeup included several genes derived from a Eurasian pig flu virus. No one was immune and so for the next 19 months the virus, officially named pandemic $H1N1/09$ virus, encircled the globe. By the end of the pandemic, the virus had infected between 11 and 21 per cent of the world's population, with around 300,000 being killed, frequently by pneumonia. Severe infections were noted in pregnant women,

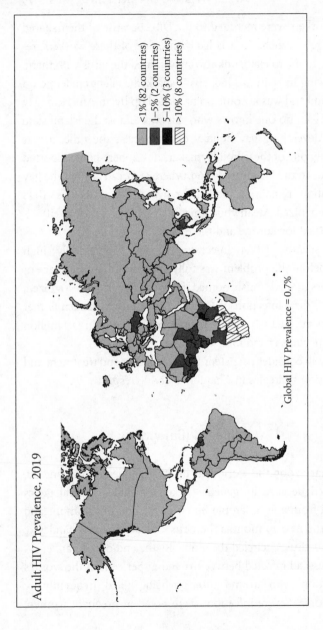

Figure 17. Global incidence of HIV-1 infections.

YOU CAN GET AIDS FROM YOUR PARTNERS' EX-PARTNERS' EX-PARTNERS' EX-PARTNERS' EX-PARTNERS' EX-PARTNERS' EX-PARTNERS' EX-PARTNERS' EX-PARTNERS' EX-PARTNERS' EX-PARTNERS' EX-PARTNERS' EX-PARTNERS' EX-PARTNERS' EX-PARTNERS' EX-PARTNERS' PARTNERS.

Find out what to do now. Call the toll-free Texas AIDSLINE. All calls are anonymous.

1-800-299-AIDS

TEXAS DEPARTMENT OF HEALTH

Figure 18. HIV poster.

but unlike seasonal flu epidemics, H1N1/09 did not specifically target the elderly, the peak age range for deaths being 50–64 years (Figure 19). This is probably because of lingering immunological memory against the H1N1 Spanish flu in the elderly, as this virus strain continued to circulate in the population for several years after the 1918–19 pandemic.

Many countries had stockpiled PPE and anti-flu drugs for just this eventuality, but in the event the infection turned out to be relatively mild, rendering these precautionary measures mainly superfluous. This was fortunate, but it had a serious knock-on effect. As a result, some governments in the West formed the impression that scientific advisors exaggerated the risk of pandemics, and this influenced their lack of preparedness for the onslaught of the COVID-19 pandemic in 2020.

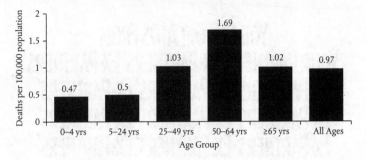

Figure 19. Chart showing age-related deaths from H1N1 swine flu during the 2009 pandemic.

Historical aspects of flu

Flu is a zoonotic infection that humans acquire from birds, and it is regarded as an emerging infection because from time to time entirely new flu virus strains with the potential to cause a pandemic jump from birds. However, human flu is not new; it has caused epidemics and pandemics for several centuries.

The term 'influenza', meaning 'influence', was coined by the Italians in the fifteenth century in the belief that flu was caused by some malevolent supernatural influence. One of the earliest documented flu pandemics in the UK was in 1562, when it was called 'the newe acquayntance'. Thomas Randolph, English ambassador at the court of Mary Queen of Scots in Edinburgh, wrote to Sir William Cecil in London:

> May it please your Honer, immediately upon the Queene's arivall here, she fell acquainted with a new disease that is common in this towne, called here the newe acqayntance, which passed also throughe her whole courte neither sparinge lordes, ladies or damoysells, not so much as ether Frenche or English.[5]

In describing the symptoms he wrote:

> It ys a plague in their heades that have yt, and a sorenes in their stomackes, with a great coughe, that remayneth with some longer, with others shorter tyme, as yt findeth apte bodies for the nature of the disease . . . There was no appearance of danger nor manie that die of the disease, excepte some olde folkes.

The human flu virus was discovered in 1933, but uncovering its origin took a while longer. In 1955 Werner Schäfer, at the Max Planck Institut für Virusforschung in Tubingen, Germany, was working on fowl plague, which regularly devastated domestic flocks, when he found that the fowl plague was caused by none other than the bird flu virus, a very close relative of the human flu virus. Schäfer suggested that under certain conditions flu viruses from different species could undergo a process of gene swapping, or 'genetic reassortment', which then enables them to infect another host species and '. . . in this way a new type of [human] influenza agent might develop from fowl plague and vice versa'.[6] This was exactly right, and explained why flu epidemics and pandemics were so unpredictable in size, severity, and periodicity.

Outbreaks of flu occur virtually every winter, while full-blown epidemics appear on average every 8–10 years. During an epidemic up to 10 per cent of the population is infected and on a worldwide scale the average death toll is estimated at around 400,000. Flu pandemics occur less frequently, however, and are much more serious than epidemics. In the hundred or so years since the devastating 1918 Spanish flu pandemic described shortly, there have been four more flu pandemics—in 1957, 1968, 1977, and 2009—and in each the virus spread round the world like a Mexican wave.

Antigenic drift and shift

The pattern of frequent flu epidemics interspersed with rarer but more devastating pandemics is down to the virus's ability to change its genetic make-up. Flu virus has an RNA genome that mutates rapidly, so that virus circulating in a community on average collects two or three changes in its protein sequence every year. This is called 'antigenic drift', because the virus slowly drifts away from the parent stock and, by the same token, away from our immunity built up by previous exposure. Eventually, the altered virus differs sufficiently to infect people who are immune to its parent, and this change heralds another epidemic.

Flu virus particles bear a fringe of spikes protruding from their surface (see Figure 4(c)), which consist of the two proteins known as haemagglutinin (H) and neuraminidase (N). These are the receptor molecules that allow the virus to attach to and gain entry to host cells. Neutralizing antibodies against H and N proteins prevent the virus from binding to and entering cells, thereby rendering a person immune. In molecular terms, antigenic drift is the slow and cumulative minor alterations in the structure of H and N which, when sufficiently different to allow the virus to dodge our defences, causes an epidemic. Nevertheless, some immunity to the virus remains, ensuring that antigenic drift alone does not cause a pandemic.

A flu pandemic occurs when a major change in the genetic makeup of the virus produces an entirely new virus strain to which the majority of the world's population is susceptible. This ability for sudden large-scale change is called 'antigenic shift'. It occurs at about 10–40-year intervals, and is unique to flu viruses.

Because H and N proteins are of major importance in immunity against flu, different flu virus strains are named according to the variants of H and N genes they carry. In all there are 16 different H and nine different N genes, and those carried by the virus strain that caused the 1918 Spanish flu pandemic were retro-spectively named H1 and N1. Following this pandemic, the same virus strain circulated in the community with minor drift until 1957 when H2N2, Asian flu, appeared. In 1968 this was replaced by H3N2, Hong Kong flu, and in the 1976–7 Russian flu a 1950s version of H1N1 reappeared. Each one of these antigenic shifts caused a flu pandemic, the most recent being the 2009 swine flu pandemic caused by an altered version of H1N1 virus with genes other than H and N acquired from North American and Eurasian pig flu viruses. Most epidemic and pandemic flu strains arise in East Asia, particularly rural Southern China. This is probably because there are more water fowl (mainly ducks) and pigs (also susceptible to flu, see the next section) kept under cramped con-ditions in that region than anywhere else in the world.

How does antigenic shift occur?

Flu virus RNA is segmented into eight separate genes. This means that if two different strains of flu virus infect the same cell, the new viruses produced may contain a mixture of the two parents' genes. When this reassortment involves H or N genes, the resulting virus may be sufficiently different from the cur-rently circulating strain for most people to be non-immune. This is the recipe for a pandemic.

Genetic reassortment of flu virus genes generally occurs in birds because they carry viruses with all 16 H and 9 N genes in almost any possible combination as a harmless, lifelong infec-tion. So birds act as a reservoir of different virus strains that from time to time spill over into humans. But only viruses carrying a few of the 16 H genes can infect human cells directly. So for a successful jump of reassorted bird viruses to humans an intermediate host is sometimes required. This is where pigs come in, since they are very susceptible to infection with both bird and human flu strains. Recombination between bird and human flu viruses can occur in pigs if they are infected with both virus strains at the same time. This may sound so complicated that in reality it would never happen, but it does. A few times every century a virus with genetic material that has been reassorted in a bird and mixed with human flu genes in a pig emerges that can infect humans and cause a pandemic. The 1957 Asian flu was caused by a human virus with three bird virus genes, and Hong Kong flu virus, which caused the 1968 pandemic, contained two bird flu virus genes.

The 1918 Spanish flu pandemic

Everyone agrees that the 1918 Spanish flu pandemic, occurring just as the First World War was ending, was exceptional. With a ferocity comparable to that of COVID-19, the virus swept around the world, reaching all the remote regions including the most isolated island communities. Around one third of the world's population was infected, and one in twenty died. The estimated death toll was over 50 million, with most fatalities among young children and young adults. This compares with a death rate of around one per thousand in other flu pandemics, where most fatalities are among the very young and very old (Figure 20).

The first cases of Spanish flu were officially reported from near the western war front in Spain—hence the name 'Spanish' flu. But earlier in 1918 there was an outbreak of milder flu in North America and this may have travelled to Europe along with American troops. However, despite the lack of supporting evidence, many scientists still believe it likely that the flu virus which caused the 1918 pandemic, like most of those which followed, arose in East Asia.

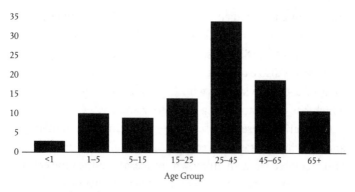

Figure 20. Chart showing age-related deaths from H1N1 flu during the 1918 Spanish flu pandemic.

In theory, the unusually high mortality during the 1918 pandemic could have been due either to the virus itself being excessively virulent or to the unprecedented circumstances of the time. Europe had been at war for four years, during which time troops lived in overcrowded, extremely unhygienic conditions, and large groups of them moved rapidly from place to place. These are exactly the conditions which encourage spread of airborne viruses like flu. In addition, many were stressed, exhausted, and poorly nourished—just the people who succumb most easily to infections. But war conditions alone, however terrible, cannot explain the severity of the pandemic, because even in parts of the world untouched by the ravages of war the death toll was unusually high. So could the virus itself have been particularly virulent?

If similar catastrophes are to be prevented in the future, it is important to answer this question, but doing so required access to the 1918 Spanish flu's RNA genome. In 2005, scientists announced that they had sequenced the complete genome of the 1918 flu virus, a remarkable feat that only became possible after they succeeded in extracting the genome from the body of an

Inuit woman who died of the disease and was buried at the remote Brevig Mission, Alaska. Here, the permafrost preserved the viral RNA for 80 years just as if it had been in a deep freeze. Researchers also extracted identical flu RNA sequences from the lung of a US serviceman who had succumbed to Spanish flu, which had been stored in a laboratory archive.[7]

Comparing these RNA sequences with common human, pig, and bird flu viruses showed that Spanish flu was an avian flu virus acquired from an unknown source, and at the time entirely new to humans. Scientists identified several mutations unique to the pandemic strain, including one that prevents infected cells from producing interferon, an important cytokine, needed to kickstart the body's immune response. This gives the virus a head start, enabling its rapid spread throughout the lungs, where it damages the tissue. The body then responds with a massive and inappropriate release of other cytokines called a 'cytokine storm', which causes severe injury to the lungs. The resulting outpouring of fluid fills the airspaces, leaving the sufferers unable to absorb vital oxygen and drowning in their own body fluids. This dreadful scenario usually proves lethal.

H5N1 avian flu

In May 1997, a three-year-old boy was admitted to hospital in Hong Kong with an acute respiratory illness and a high fever. A few days later he was dead, but not before the offending flu virus was isolated from his respiratory tract. This episode could have passed unnoticed but for the fact that scientists at the Hong Kong Department of Health could not identify the strain of the flu virus. They sent it to the WHO Collaborating Centres for Flu Surveillance where the answer was unanimous: H5N1—an entirely new flu strain to infect humans, but identical to the one which had killed 4,500 chickens on three Hong Kong farms earlier that year. In this first human outbreak there were 18 cases and six deaths, although the virus did not spread from person to person. After culling all the local poultry the outbreak was stamped out, but in December 1997 hundreds of chickens in a Hong Kong market and on a farm near the Chinese border caught the virus and died. And this time 17 humans, all poultry handlers, caught H5N1 flu and four of them died.

'World facing killer bug', the newspaper headlines screamed, and immediately dubbed the offending virus 'bird flu'. With panic mounting among its local inhabitants, and a rapidly falling tourist count, the Hong Kong Government decided to act. All Hong Kong's 1.2 million chickens from 200 farms and 1,000 trading posts were slaughtered within 24 hours. Chicken imports from China were banned, which left Hong Kong devoid of fresh chickens for its approaching traditional New Year banquets and, presumably, full of irate chicken farmers, traders, and chefs.

The officially named H5N1 Asian flu virus spread rapidly among poultry, causing a panzooic (a pandemic in animals), affecting more than 50 countries. It is currently endemic in flocks in China, Thailand, Bangladesh, Egypt, Indonesia, and Vietnam. Scientists analysing the viral genes discovered that it contains the same mutation as that found in the 1918 pandemic virus, which can trigger a cytokine storm. This is undoubtedly why the virus is so lethal in chickens, killing around 95 per cent of infected birds.

Fortunately, at the time of writing, H5N1 Asian flu virus has not adapted to spread among humans. Nevertheless, the WHO reports 862 human cases since 2003, with an approximate death rate of 60 per cent. In the meantime, several other novel flu viruses have emerged, some of which can infect humans. Without a vaccine to protect us from these novel viruses, the policy for preventing another epidemic or pandemic is to cull infected flocks and continue close surveillance to monitor their spread and detect any changes in human infectivity. The prevention and treatment of flu are discussed in Chapter 8.

Severe acute respiratory syndrome coronavirus-2 (SARS-CoV-2)

In December 2019 a new coronavirus that infects humans emerged in Wuhan, the capital city of Hubei province, China. The virus, SARS-CoV-2, caused a local outbreak of a severe respiratory disease—coronavirus disease-19 (COVID-19). When this spread more widely, Wuhan and other cities in Hubei province were sealed off, with schools closed and residents confined to their homes. A travel ban was implemented, but unfortunately the outbreak coincided with Chinese New Year celebrations and millions of people had already left the province, some carrying the virus with them. The world was alerted to the outbreak when just 59 COVID-19 cases had occurred in Wuhan—but it was already too late.

COVID-19 spread throughout China, into neighbouring countries and then around the globe with amazing speed. Similar lockdowns and travel bans were imposed by other countries in an attempt to prevent virus importation and spread. But in Europe, the US, and many other areas, it was a case of too little too late; the virus had already sneaked in.

COVID-19 first reached Washington, US, on 15 January 2020; Bavaria in Germany on 20 January 2020; and Lombardy in Italy on 28 January 2020—all carried by travellers direct from China. From Italy, the epicentre of the European epidemic, the imported virus spread widely, initially seeded by holidaying skiers returning home. It then jumped direct from Italy to New York on 12 February 2020, initiating the city's largest transmission cluster.[8] The WHO declared a public health emergency of international concern on 30 January 2020 and upgraded this to a pandemic on 11 March 2020.

COVID-19 spreads between people via mucus droplets in exhaled air generated by coughs and sneezes and also talking and singing. The virus infects through the mouth and nose, and on occasions can also infect by contamination of the eyes. In order to infect a cell, SARS-CoV-2 uses its spike protein, which, as the name suggests, forms the spikes on the surface of the virus particle (see Figure 4(b) in Chapter 1). The spike protein docks with a cell receptor called ACE-2 (angiotensin-converting enzyme-2), which is very widespread in the body, expressed on cells lining blood vessels and cells in the respiratory tract, kidney, and gut. This cell receptor distribution determines the spread of the virus within the body and therefore the symptoms of COVID-19.

On entering the body, the virus infects cells of the respiratory tract, and 5–7 days later the infected person becomes *infectious*, even if, as in around 20 per cent of cases, the infection is asymptomatic. The most common presenting symptoms of COVID-19 are fever, cough, shortness of breath, and loss of taste and smell. These symptoms may be mild, but, particularly in the elderly and those with chronic health problems, can be life threatening. The virus spreads throughout the body, affecting many organs. In particular lung infection may progress to severe respiratory disease often caused by a cytokine storm reminiscent of 1918 Spanish flu. This requires hospital admission, and may necessitate intensive care and mechanical ventilation. A blood clotting disorder may occur in those with severe disease, and this can compromise the function of vital organs such as the brain, lungs, and kidneys.

Death rates from COVID-19 increase with age—from less than 0.1 per cent in people up to around the age of 50 to as high as 25 per cent in those over 85 years old—and are higher in males than females, in hospital staff, and in ethnic minority groups. The overall death rate is around 1 per cent (Figure 21).

When COVID-19 first strikes a population no one is immune so the virus spreads rapidly. Within 55 days of its discovery 100,000 COVID-19 cases had been reported; 20 days later this figure had reached 500,000; and a year later there had been over 100 million reported cases and 2.5 million deaths worldwide—and this takes no account of undiagnosed or asymptomatic infections. The reproductive number (R0) for the virus—which, recall, is the average number of cases spawned by one infected person at the beginning of an epidemic—is 2–3, but

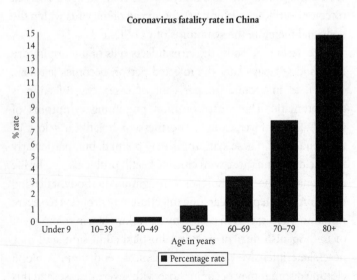

Figure 21. Chart showing age-related deaths from COVID-19. Data provided in February 2020.

there have also been reports of 'super-spreaders' such as a cluster of 100 cases infected by a single individual during an international business conference in Boston, US, in early March 2020.[9] And, early in the pandemic, cruise ships were a hotspot for virus transmission. *The Diamond Princess* with its 3,711 passengers and crew provided a rare opportunity to study a microcosm of COVID-19 spread among its mainly elderly passengers in a confined space.[10] On 1 February 2020, a passenger who had disembarked a few days earlier tested positive for COVID-19. So when the ship arrived in Japanese waters on 3 February it was immediately quarantined. A month later, 700 infections had been recorded on board, of which 18 per cent had no symptoms, and the fatality rate was 1.1 per cent.[11]

As the first wave of COVID-19 hit Europe and the Americas in March 2020, health services were overwhelmed by an explosion of severe cases. In many countries staff numbers proved inadequate; vital PPE, ventilators, and piped oxygen were in short supply; and diagnostic tests were either unavailable or too costly. Countries and States responded to the crisis in a variety of ways. In general, those that had experienced SARS back in 2003— including China, Vietnam, South Korea, Hong Kong, Singapore, Taiwan, and Japan—were best prepared. They had stockpiled PPE and had track, trace, and isolate systems rapidly up and running. This pre-prepared and decisive response avoided the sharply rising cases and deaths seen elsewhere. Certain island states like New Zealand also fared well by using rigorous travel bans to avoid imported cases, thereby keeping control over the track, trace, and isolate policy.

In the UK, like most of Europe, track, trace, and isolate failed to keep up with the rapidly rising COVID-19 cases, and this left the government relying on traditional public health measures such as hand washing, wearing masks, and social distancing, with public information posters encouraging these measures. When this also failed to curtail virus spread, full lockdown was declared on 16 March 2020.

Initially, the Swedish government took a more relaxed approach to the crisis, with no lockdown or closures of schools or of the hospitality industry. This was a deliberate attempt to develop herd immunity—that is, to raise the level of immunity in the population until it prevents the virus from circulating. Scientists calculated that this required 67 per cent of the population to be immune to COVID-19. However, in Sweden this figure was not reached before rising case numbers prompted the implementation of restrictions.

The role of epidemiologists and mathematical modellers in predicting virus spread and case numbers under a variety of intervention methods was essential in informing government decisions, particularly on when to impose, and when to lift, lockdown restrictions. Key to monitoring the pandemic was the value of R. If this continued to rise above one then extra control measures were required, and only when its value was consistently well below one was it safe to ease these restrictions. In the UK, this was eventually the case in May 2020, when easing of lockdown began. But although the summer months gave some respite, a second wave of COVID-19 necessitating another lockdown hit the northern hemisphere in 2020 (see for example the UK, Figure 22).

The emergence of SARS-CoV-2

We know that SARS-CoV-2 emerged in Wuhan, a large city of 11 million inhabitants, but exactly how it emerged remains a mystery. Early on, the finger was pointed at the seafood and live

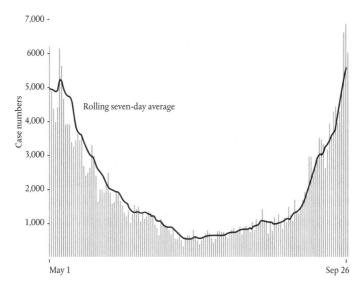

Figure 22. U-shaped curve showing the two waves of COVID-19, May–September 2020, in the UK.

animal market in Wuhan. This was based on the fact that several market workers and visitors were among the first recognized COVID-19 cases. Most scientists think that SARS-CoV-2 is derived from a bat virus since these animals carry many coronaviruses, one of which has 96 per cent identity with SARS-CoV-2. Nevertheless, a large variety of animals are sold in the market, among them pangolins, which also carry coronaviruses very similar to the pandemic strain. Some suggest that this animal may have acted as an intermediate host between the long-term carrier, most likely an as yet unidentified species of bat, and humans. Other suggested scenarios take into account the close proximity to the epicentre of the pandemic of the Wuhan Institute of Virology, where studies on emerging bat viruses are ongoing. Although, a

WHO team of international experts which visited Wuhan in early 2021 to uncover the true origin of SARS-CoV-2 ruled out a role for the Institute in the emergence of the virus, investigations are currently in progress regarding this option.

Further discoveries

The study of SARS-CoV-2 and COVID-19 is a fast-moving field, and recent discoveries have added to the unprecedented and complex nature of the pandemic. First, it soon became apparent that those recovering from severe COVID-19 may suffer debilitating sequelae. Now called 'long COVID', the syndrome comprises a multitude of symptoms including continued cough and shortness of breath as well as fatigue; head, chest, and joint pains; depression; and insomnia. This affects around 5 per cent of COVID-19 sufferers of all ages, although it is most common in the elderly, with symptoms persisting for several months. There are many unknowns about this ailment, and so far no specific treatment is available.

Second, on occasions, SARS-CoV-2 has jumped from humans to domestic cats and dogs, zoo animals such as lions and tigers, and a variety of primates and bats, but no reciprocal animal to human virus transmission had been recorded. That is why reports of a mutated form of SARS-CoV-2 in farmed mink in Denmark jumping back to humans in November 2020 came as a shock to scientists and the public alike. This scenario is called 'reverse spillover' and 'spillback'. So, while the first spillover involved SARS-CoV-2 jumping to humans from an animal source in December 2019, the reverse spillover infected farmed mink from humans, and now there is evidence of spillback of virus mutants from SARS-CoV-2-infected mink to humans. This necessitated culling of 17 million mink as a safety precaution.

Third, SARS-CoV-2 mutants or variants appeared in the UK (alpha variant), South Africa (beta variant), and Brazil (gamma variant), during January 2021, and sometime later in India (delta variant) and elsewhere, which have all mutated to a form that spreads more easily between humans. These variants will inevitably replace the original virus strain throughout the world, and so a major concern is whether the first tranche of vaccines will protect against these and other new variants. This topic is discussed in Chapter 8.

COVID-19 research

In mid-January 2020, Chinese scientists shared the first genetic sequence of SARS-CoV-2 with research groups globally. This allowed diagnostic tests to be developed early in the pandemic and now rapid tests are being used at airports, in educational establishments, and in healthcare settings, helping to curtail virus spread. Meanwhile, scientists around the world used the gene sequence to design vaccines against COVID-19; more than 100 were in production within a few months of the outbreak. Similarly, more than 1,000 drug trials were initiated looking at the effectiveness of known antiviral drugs as well as Chinese medicines. Unfortunately, only one antiviral, remdesivir, had been authorized for use by the end of 2020. The steroid dexamethasone improves survival of intensive care cases, and two drugs used for rheumatoid arthritis —tocilizumab and sarilumab—have anti-inflammatory properties that have some benefit in counteracting the cytokine storm.

Finally, it is of interest to compare the COVID-19 pandemic with other similar outbreaks, particularly those caused by SARS and 1918 Spanish flu, to address the issue of why Spanish flu and COVID-19 caused pandemics while SARS did not. The similarities and differences between these viruses and the diseases they cause are listed in Table 3. All are airborne, zoonotic viruses that

Table 3. Comparison of features of Spanish flu, SARS, and COVID-19.

Criterion	Spanish flu	SARS	COVID-19
Place and year of origin	Spain, 1918	China, 2002	China, 2019
Primary host	Birds	Bats	? Bats
Intermediate host	No	No	? Pangolin
Virus family	Orthomyxovirus	Coronavirus	Coronavirus
Spread	Airborne	Airborne	Airborne
Cell receptor	Sialic acid	ACE-2[1]	ACE-2
Incubation period	2–3 days	2–7 days	4–12 days
Super-spreaders	Not known	Yes	Yes
At-risk groups	Young children, adults 20–40 yrs	Elderly and chronically sick	Elderly and chronically sick
Mortality %	10–20	30–40	approx. 1
Asymptomatic infections	Yes, but level not known	0	20%
Cytokine storm	Yes	Yes	Yes
Ro	2–3	2–3	1.8–3.6

[1] ACE-2 = Angiotensin-converting enzyme 2.

may result in severe lung disease due to a cytokine storm which is often lethal. However, generally it is not the disease a virus causes that determines its ability to spawn a pandemic, but rather factors relating to its ability to infect and spread between hosts. Ro is an important indicator of virus spread at the beginning of an epidemic, yet the differences in outcome between the 1918 flu, SARS, and COVID-19 are not reflected in great differences in the respective values of Ro, which are all between 2.0 and 3.5.

Airborne spread is a highly efficient way of transporting viruses between hosts, but there are important differences in the method by which this is accomplished. For example, sneezing generates a fine aerosol spray of virus-containing mucus droplets from the upper respiratory tract, one sneeze producing around 40,000 droplets expelled at a speed of 200 mph. These droplets, as well as those produced by talking and singing, will remain airborne for longer, and travel further, than heavier mucus droplets generated in the lungs and released by coughing. This has been suggested as an explanation for the difference in spread between SARS on one hand and COVID-19 and flu on the other.

So although the viruses causing SARS and COVID-19 share a cell receptor, ACE-2, this is widespread within the respiratory tract, so does not limit virus production to a particular site. But because SARS causes severe lung disease in 40 per cent of cases, while this figure for COVID-19 is just 1 per cent, it is likely that SARS virus is mainly produced in the lungs and shed in heavy droplets, while during COVID-19 and flu the viruses are usually generated in the upper respiratory tract and therefore shed in lighter droplets. This difference in site of virus production would mainly restrict the spread of SARS to close contacts.

Another stark difference between the three viruses is their ability to cause a silent infection during which the host is shedding virus. This is known to be common in flu epidemics and pandemics (although the exact level of asymptomatic infections for Spanish flu is not known), and reaches up to 20 per cent for COVID-19 infections. In contrast, silent infection does not occur with SARS. In addition, COVID-19 sufferers shed the virus for a few days before developing symptoms, and super-spreading events are well documented. So while in the 2002–3 outbreak contacts of SARS cases could be traced and isolated before they became infectious, both COVID-19 and Spanish flu viruses can more easily circumvent any preventative measures short of complete lockdown, and no doubt this was a major contributing factor to their worldwide spread.

Clearly, developing an effective vaccine against COVID-19 had to be top priority, as this preventative measure is the only way to stop a pandemic from continuing until its spread is eventually checked by herd immunity. This topic is discussed in Chapter 8.

5

PAST EMERGING VIRUSES

The emerging viruses discussed in Chapters 3 and 4 are new to humans, but viruses have been jumping to us ever since we evolved from our ape ancestors. While the source of past emerging viruses, like our most recent acquisitions Ebola, SARS, and COVID-19, has usually been zoonotic, the reasons for their jump to humans have changed with time.

Today, it is close contact with wild animals that gives viruses most opportunities to cross species. And then our increasing population and ability to travel long distances rapidly allows them to spread successfully among us. But for our ancestors, it was their change in lifestyle from hunter-gatherer to farmer some 10,000 years ago that triggered an onslaught of new, emerging infectious diseases. In this chapter we look at how these ancient afflictions have evolved over the intervening 10,000 years with a view to understanding how diseases like COVID-19 might evolve over time. The examples used here are smallpox and measles, both highly infectious, lethal airborne viruses, as well as polio, spread by faecal–oral contamination.

The advent of agriculture embraced two fundamental lifestyle changes that together had a profound effect on our ancestors' susceptibility to microbes. By all accounts, hunter-gatherers were fairly free of infectious diseases because of the small size of their groups and their free-ranging lifestyle. But the switch from the nomadic life to living in fixed communities, along with the change from hunting animals to their domestication, encouraged spillover of new viruses as well as other types of microbes. So, the early farmers' emerging viruses jumped from the animals they domesticated, while the cramped conditions of life in villages, and later towns, gave these viruses the opportunity to thrive in their new human host. This was the beginning of many of our acute childhood infections, so-called crowd diseases.

The construction of evolutionary trees of virus families can often tell us when, and from where, human infections arose. Smallpox virus, for example, is most closely related to gerbil pox and camel pox viruses, and is thought to have crossed species from gerbils to camels and humans around 5,000–10,000 years ago. This was probably the first of the known early human epidemic viruses to be acquired; somewhat later measles evolved in humans from rinderpest virus of cattle, with many other microbes following suit.

Just like Ebola, SARS, and MERS, these viruses initially jumped from their primary host at the start of each outbreak and died out again when there were no more susceptible people to infect (albeit without any human intervention to curtail virus spread). As communities grew, and communication between them increased, they could support larger and more widespread epidemics, until eventually the viruses could be sustained permanently among humans. At this point the viruses broke their ties with their animal hosts and, now resembling HIV-1 and COVID-19, evolved independently in humans.

This transition probably began around 1000 BC when the flourishing civilizations in the fertile valleys of the Nile in

Egypt and the Ganges in India were sufficiently populous, and their cities crowded enough, to sustain the circulation of these emerging viruses independent of their primary hosts. And with no genetic resistance to help fight these new plagues, they caused much more severe disease than the related childhood infections we experienced after thousands of years of coevolution.

Once free of their original animal hosts, emerging viruses like smallpox and measles had no backup reservoir. And since they induce lifelong immunity after a single infection, immunity built up during an epidemic until, at a certain level, the virus could no longer circulate in the community. At this point, although not everyone would have been infected, the whole population would be protected—herd immunity. The level of acquired immunity needed to reach herd immunity varies between viruses depending on their ability to spread, that is, their R0 number. For measles, the most infectious of viruses (R0 12–18), a level of 90–95 per cent of the population with immunity is required, while for smallpox (R0 3.5–6.0) it was around 80 per cent, and for flu (R0 1.4–2.8) 33–44 per cent.

In practice, herd immunity means that viruses which cannot infect the same individual twice, and do not establish a persistent infection, require large populations of people living in close contact to survive. Only then can repeated cycles of infection occur in the constantly renewing pool of susceptible children born after the last epidemic.

Smallpox

In its heyday smallpox was a major killer. This highly infectious, airborne virus caused a much-feared disease that began with the non-specific symptoms of fever, sore throat, headache, and general aches and pains that might herald many infectious diseases, but then on the fourth day the typical rash appeared, confirming the dreaded diagnosis. Around 30 per cent of those infected died of the acute disease, and many others were left blind or disfigured.

Although there are no written records of smallpox from the twelfth century BC, mummified bodies from Egypt, particularly that of King Ramesses V, who died as a young man, show typical smallpox pustules (Figure 23). The virus probably spread from

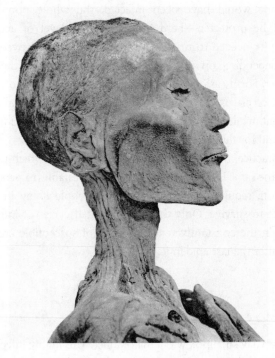

Figure 23. Mummy of Ramesses V showing lesions on the face suggestive of smallpox.

its strongholds in Egypt and India north and east to China, where the first written records of the disease date from AD 340.

From these early beginnings, smallpox, along with other emerging microbes, expanded its range by travelling with traders, armies, and colonizers, always taking advantage of virgin populations to infect, and causing devastating epidemics wherever it went. Smallpox travelled ancient trade routes like the Burma Road and the Silk Route, and arrived in Europe with Islamic invaders and returning Crusaders. By the twelfth century, it was rife in most of Europe, Asia, and North Africa, where it established the pattern of periodic explosive epidemics causing over 400,000 deaths every year in Western Europe.

Between the fifteenth and eighteenth centuries smallpox was transported from the Old World to the Americas, the whole of Africa, Australia, and New Zealand by early settlers and invaders. In each of these territories the indigenous peoples suffered terribly because of their total lack of genetic resistance to the new plague. In some places the disease killed up to half the population, clearly having a significant influence on local history. The defeat of the Aztecs of South America by Spanish invaders is a case in point.

Hernando Cortez had under 600 troops when he invaded Mexico in 1520, and he was being soundly beaten until he was joined by a small relief force which happened to include someone incubating smallpox. Because the Spanish were probably mostly immune to smallpox, having suffered childhood infection, they were unaffected. But the virus was entirely new to the Aztecs. Almost every one of them came down with smallpox and it killed around a third of the population, so allowing Cortez an easy victory.

At the same time back in the Old World, smallpox epidemics continued to escalate as populations of towns and cities grew and became more crowded. But although smallpox spread most efficiently among the cramped dwellings of poor city dwellers, the virus did not spare the elite or the wealthy.

At its peak, the virus accounted for over 5 per cent of all deaths in London. The Tudor Queen Elizabeth I suffered a from severe bout of the disease in 1562, and the House of Stuart, which succeeded the Tudors to the monarchy, lost several members to smallpox, and was finally wiped out when its last heir, Queen Anne's only son, Prince William, died of the disease. Within eighty years of these tragic events, many European royals succumbed, including Emperor Joseph I of Austria, King Luis I of Spain, King Louis XV of France, Queen Ulrika Eleanora of Sweden, and Tsar Peter II of Russia.

None of the early treatments recommended for smallpox sufferers improved the outcome, and they likely caused unnecessary stress. The 'heat treatment' was conceived in the tenth century by Al-Razi, the famous Persian physician who was the first to distinguish smallpox from measles. He believed that smallpox was caused by fermentation of the blood. This, he thought, produced humours that escaped from the body through skin pores. To assist this process he advocated heat, with the sufferer sitting in a sealed room in front of a blazing fire to sweat out the ill humours. This practice was embellished in twelfth-century Europe by the 'red treatment', which had the unfortunate patient dressed in red clothes, wrapped in red blankets, confined to a red room and attended by red-dressed carers. This bizarre concept possibly originated in Japan where it was believed that a red coloration could expel illness and demons. The practice continued until the seventeenth century, when the English physician Sir Thomas Sydenham noted that patients receiving the red treatment had a worse outcome than those unable to afford it. Sydenham advocated no treatment for mild cases and 'cooling treatment' for the more severely affected. He threw open the windows to expel the evil humours and in doing so improved survival. Nevertheless, the virus continued its killing spree un- abated until the first smallpox vaccine, famously produced by Edward Jenner, was introduced in 1798.

Measles

Measles is one of the most infectious of viruses, and although today it mainly causes non-life-threatening disease in healthy populations, it is still a killer for the weak and undernourished. The WHO reports 140,000 global measles deaths annually, mostly of children under five years old in the poorest countries.

Measles was highly lethal when first introduced to societies with no previous experience of the virus. Serial measles epidem- ics were instrumental in extinguishing native island populations, including that of the remote islands of Tierra del Fuego, situated at the southernmost tip of South America. Here, Scottish mis- sionaries, Thomas and Mary Bridges established an outpost in 1871. All went well until a visiting supply ship brought a mystery illness, first diagnosed by the doctor as typhoid pneumonia, but soon recognized as measles by Mrs Bridges. The native popula- tion suffered severely. More than half of those infected died of the acute disease, and many of the survivors never regained their strength and died within two years of the initial infection. In contrast, European children at the mission only suffered a mild illness, and European adults, who had all had measles in their youth, were unaffected.

E. Lucas Bridges, son of Thomas and Mary, wrote a first-hand account of the effect that measles had on the Yahgan tribe:

The natives went down with the fever one after the other. In a few days they were dying at such a rate that it was impossible to dig graves fast enough. In outlying districts the dead were merely put outside the wigwams or, when the other occupants had the strength, carried or dragged to the nearest bushes.[1]

He goes on to speculate (correctly) on the reasons for the racial differences in the outcome of infection:

It must be that our ancestors, for generations past, have suffered from periodic epidemics of measles, and we, in consequence, have gained a certain degree of immunity from it. On the other hand, the Yahgans, though incredibly strong, and able to face cold and hardship of every kind and to recover almost miraculously from serious wounds, had never had to face this evil thing, and therefore lacked the stamina to withstand it.

Cyclical epidemics of airborne viruses like measles became routine over time, but the reason for these regular cycles of infection that mainly target young children was not understood until 1846, when a now famous measles epidemic broke out in the community living on the remote Faroe Islands.

Measles had not visited the Faroe islands for 65 years, when a carpenter arrived from Denmark eight days after visiting a sick friend. He rapidly developed measles, and in the following six months 6,000 of the 7,782 islanders also caught the disease. Peter Panum, a young Danish medical health officer, was sent to the island to investigate, and spent the next five months unravelling the epidemiology of the outbreak.

By finding that none of the elderly residents who had measles in the previous epidemic of 1781 caught the infection, Panum established that immunity following natural measles infection is lifelong. Then, by carefully noting who his patients came into contact with, he showed that the incubation period is 13–14 days.

These observations finally explained why measles and other acute infectious diseases usually occurred in children, and why these viruses swept through communities at regular intervals. For measles, 90 per cent of children in towns and cities were immune by the age of 15, and this level of infection continued until a measles vaccine was licensed for use in 1968. And after vaccines to mumps and rubella were developed, the trivalent MMR became available in 1988.

Poliomyelitis

A young man with drop foot, beautifully depicted on a funerary stele dating from the eighteenth Egyptian dynasty (1580–1350 BC), is perhaps the earliest record of a virus infection and shows us that paralytic polio is an ancient affliction (Figure 24). However, the disease only became a significant public health problem in the twentieth century, and even then mainly in western societies.

The epidemiology of paralytic polio is counterintuitive; this much-feared illness specifically flourished in countries with *high* standards of living, so confounding the expected pattern of infectious diseases. Polio epidemics usually struck during the summer months, mostly affecting children of affluent families. The illness started abruptly with headache, fever, vomiting, and a stiff neck, as the virus spread through the body. It then homed in on the central nervous system, particularly the spinal cord, damaging the nerves to such an extent that a child who was perfectly healthy one week could be permanently paralysed the next. Anything from part of a single muscle to virtually the whole body could be affected. When the virus attacked the nerve supply to the respiratory muscles victims were condemned to life inside an iron lung (an early version of a ventilator). Many

Figure 24. Young man with drop foot, probably caused by polio, from a carving in an Egyptian tomb of the eighteenth dynasty.

well-known people suffered from polio, among them the author Sir Walter Scott who recovered but for a limp, and US President Franklin Roosevelt who, from the age of 40, was permanently paralysed below the chest (Figure 25).

In Europe and North America, polio epidemics reached their peak in the 1940s and 1950s and then quickly disappeared as the first vaccine was rolled out from 1955 onwards. At the same time, polio began to emerge as a major health problem in developing nations. These remarkable changes in disease pattern in less than a hundred years were caused by alterations in our lifestyle which profoundly affected the epidemiology of the virus.

Figure 25. Photograph of President Roosevelt with polio sufferers.

The history of polio research is an unhappy one, and this goes a long way to explaining why it took nearly 50 years from the discovery of the virus in 1908 to achieve the production of a vaccine in 1955. Karl Landsteiner and Edwin Popper, working in Vienna, identified polio virus in a filtered extract from the spinal cord of a fatal case by using it to successfully transmit the disease to two Rhesus monkeys. Thereafter, polio infection of rhesus monkeys became the model for human polio in virtually all polio research for the following 25 years.

But because rhesus monkeys were so expensive, one research group dominated the field—that of Simon Flexner at the Rockefeller Institute for Medical Research, in the US. These scientists firmly believed that the natural route of polio virus infection in humans was through the nose; they assumed that virus travelled directly from this site along nerves to the brain without infecting any other tissues. However, during a polio epidemic in Sweden in 1911, Carl Kling, an assistant in the Bacteriological Institute in Stockholm, detected polio virus not only in the noses of polio sufferers, but also in their mouths and the wall and contents of their intestines. Moreover, he also found the virus in the intestines of healthy family members of sufferers and even in healthy controls among the general public. He was on the right track, but when he reported these discoveries the following year they were assumed wrong and totally ignored—until, that is, some 26 years later. The Flexner theory was eventually debunked when John Paul, a physician at Yale University, repeated Kling's findings. Finally, all the pieces of the polio puzzle began to fall into place.

Polio virus spreads by the faecal–oral route. It enters the body through the mouth and, although it does not cause overt gastroenteritis, the virus grows in the cells lining the small intestine. At this stage large amounts of virus are excreted in faeces that can survive for several weeks in water and sewage, thereby enabling passage to other victims.

Polio infection is *not* a rare event; rather it is common but usually asymptomatic, and nerve paralysis is an exceptional manifestation of this common infection. Only 0.1–1.0 per cent of infections result in paralysis, although the reason for this is unclear. Polio virus naturally thrives in overcrowded, unhygienic conditions. Consequently, until the second half of the twentieth century, the virus was endemic in developing countries and in poorer areas of industrialized countries. Here, asymptomatic infection occurred early in life, and by the age of five virtually all children were immune. In this situation the paralytic form of the disease was extremely rare. Conversely, in affluent societies of Europe and the US, where standards of hygiene were high, young children were often protected from infection. Under these conditions a sufficiently large susceptible population periodically built up, epidemic conditions prevailed, and the virus spread through the community, with a small minority of cases developing paralytic disease.

This pattern of infection is aptly called 'the iceberg phenomenon' because the majority of infections are below the level of detection, with just those at the very tip suffering from paralytic disease. In developing countries, the change from an endemic to an epidemic pattern of polio seen during the second half of the twentieth century was the inevitable consequence of rising standards of living.

In 1948 John Enders and co-workers at the Boston Children's Hospital in the US succeeded in growing polio virus in tissue culture, a prelude to vaccine development and a feat that won them the 1954 Nobel Prize in Physiology or Medicine. Introduction of polio vaccine in 1955 had a dramatic effect on epidemic polio in the industrialized world. In the US the incidence of paralytic polio immediately dropped tenfold, from the pre-vaccine level of around 20,000 cases per year. Similar case reductions followed in many other countries.

Global virus elimination

The great success of the smallpox vaccine—the first ever vac-cine, famously invented by Edward Jenner at the end of the eighteenth century—prompted the WHO to begin its highly ambitious Worldwide Smallpox Eradication Campaign in 1966. This was deemed possible because the virus had already been eradicated from the US and most of Europe by the 1940s. But as smallpox still posed a threat in these areas because of imported cases that necessitated strict surveillance and quarantine, their governments were willing to provide funding for the campaign. At the beginning of the campaign smallpox was endemic in 31 countries. In that year 131,776 cases were reported, with a 20 per cent fatality rate. Given that this probably underestimated the total number of cases a hundredfold, smallpox was obviously still a major world health problem, and governments rich and poor were united behind the campaign.

Operationally, the WHO team, headed by Don (D. A.) Hen-derson, defined a smallpox endemic area as anywhere with more than 5 cases per 100,000 population and less than 80 per cent of people vaccinated. They assigned each of these areas to one of four isolated endemic zones with little chance of cross infection between them—Brazil, Indonesia, sub-Saharan Africa, and the Indian subcontinent (Figure 26). The aim was to interrupt the spread of the virus since, because smallpox was an acute infec-tion with no animal reservoir, if they could break all chains of infection the virus could not survive. The priorities were to

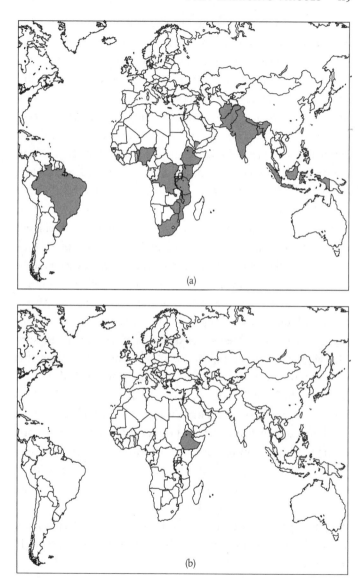

Figure 26. World maps showing smallpox endemic areas in (a) 1968 and (b) 1975.

improve vaccination rates and increase case reporting, to trace all contacts of cases, and to isolate cases and contacts. This required a massive investment; fieldworkers and laboratory staff worked hand in hand to produce an effective, heat-stable vaccine, to vaccinate in remote and sometimes reluctant communities, and to diagnose acute cases and trace all their contacts.

This strategy was so effective that by 1973 only six endemic countries remained—India, Bangladesh, Pakistan, Nepal, Botswana, and Ethiopia. The work continued unabated, and by 1975 Ethiopia was the only endemic area left. A year later the virus was gone. But before any country could be declared smallpox-free, a 'quarantine' period had to elapse during which surveillance continued as before. Monkeypox and chickenpox often look remarkably like mild smallpox, so all rumoured cases had to be checked. The following figures from India, for a single month in 1976, show the staggering extent of the operation: 668,332 villages, including 3,051,753 households, were searched by 115,347 workers, who were themselves overseen by 29,046 supervisors. The work was monitored by 8,048 assessors who found that 97 per cent of the 107,409 villages they assessed had, in fact, been searched. All this effort uncovered 41,485 cases of chickenpox but no smallpox. Thus India was declared smallpox-free.[2]

Almost unbelievably, the last two recorded cases of smallpox in the world were in Birmingham, UK. In 1978, a photographer working in the Department of Anatomy at the University of Birmingham Medical School died of the disease; one other individual caught the infection from her but recovered. Neither the photographer nor anyone else in the Anatomy Department worked with smallpox virus, but large quantities were grown in the Department of Microbiology situated in the same building on the floor below. An enquiry into the catastrophe found that the photographer was infected with the same strain of the virus being used in the smallpox laboratory, where safety procedures were reported as 'far from satisfactory'.[3] The virus probably travelled through an airduct leading from the smallpox preparation area to the floor above where the enquiry team found an ill-fitting airduct inspection panel close to a telephone that the photographer used several times a day. The report pointed the finger at smallpox researchers for breaking the safety rules, the Microbiology Department for not enforcing them, and the inspector from the Dangerous Pathogens Advisory Group for failing to realize the extent of the hazard. The head of the department at fault killed himself, so bringing the death toll from this regrettable incident to two.

Worldwide smallpox eradication was finally declared in 1980. The whole enterprise cost $312 million, of which an estimated $200 million would otherwise have been spent on routine control programmes. This incredible milestone in our triumph over lethal viruses heralded a new era in the fight against infectious diseases. When it comes to elimination, each virus has its own particular defences to be overcome; some lurk in animal hosts, some set up persistent infections which are difficult to ferret out, and for others we do not yet have a foolproof vaccine. Nevertheless, the long-term benefits of virus eradication, in financial as well as health terms, are clear, and so, although success may be several decades away, it should certainly be our long-term goal.

With this in mind, the WHO is now committed to the global elimination of polio, measles, and rubella. The programmes have already eliminated these viruses from vast areas, but sadly the goal of total eradication by 2020 was missed, mainly because of civil unrest and wars, as well as the COVID-19 pandemic. But hopefully within a decade polio, measles, and rubella, like smallpox, will be only a receding memory.

The relevance today of past emerging infections

Like our early farming ancestors, we are living through an age of increasing emerging infections—so how can their experiences inform our actions, particularly regarding pandemics such as COVID-19? Of course, the two situations are very different. Over the intervening 10,000 years the global population has mushroomed from less than 10 million to 7 billion! And travel has advanced enormously, so while we, and therefore our viruses, can circumnavigate the world's 24,902-mile (40,000-km) circumference in around two to three days, early farmers (walking at 3 miles an hour for 12 hours a day) would have taken 690 days over the same trip.

Clearly, these stark differences in interconnectedness meant that stopping the rapid spread of COVID-19 was an impossible feat right from the start; once it was out of the box there was no going back. But the principles of control are the same as ever. Our ancestors had no option but to wait for the natural brake, herd immunity, to kick in, but this might take several years, and many deaths. Likewise, genetic resistance, which would eventually have reduced the severity of an infectious disease, possibly took as long as 100–200 years of coevolution. So, accepting that COVID-19 is not going away of its own accord, and that eradication, in the short term at least, is impossible, our best option has been to aim for herd immunity.

A year into the pandemic, levels of natural immunity against COVID-19 in most countries did not exceed 20 per cent, whereas scientists reckon that this figure needs to be 67 per cent before herd immunity will kick in. So much more illness and death would have to be suffered before reaching this threshold through natural immunity. But thankfully, we had twenty-first-century science and technology on our side, and this has massively speeded up our response to the crisis.

As soon as, in January 2020, COVID-19 was recognized as a new and lethal disease, the entire viral genome was sequenced within weeks, and test kits were available shortly afterwards. Then, almost miraculously, several vaccines against the SARS-CoV-2 virus were licensed for use within a year of its emergence. These vaccines are our way out of the crisis—our pathway to herd immunity—but only if, like those against smallpox, measles, and polio, they prevent infection and reinfection, and interrupt virus spread completely. In that case aiming for herd immunity would be an achievable goal.

This is the best-case scenario, but suppose the vaccines do not produce lasting immunity; prevent disease but not infection; or, like flu vaccines, fail to protect against viral mutants. Despite the incredible success of the smallpox, polio, and MMR vaccines, they do not necessarily represent the norm. In many cases effective vaccines cannot be made—the common cold being a case in point. And since coronaviruses can cause the common cold, it is pertinent to take a look at these milder cousins of SARS-CoV-2.

Coronaviruses and the common cold

In the mid 1960s, David Tyrrell, working at the Common Cold Research Unit in the UK, discovered the first human coronavirus. Two more common cold-causing human coronaviruses came to light in the 1960s, but at the time these discoveries engendered little interest. However, by the time the latest common cold coronavirus was discovered in 2005, everything had changed. With the emergence of their more aggressive cousin, SARS-CoV, in 2003, virologists made up for this oversight. We now know that these viruses can cause more severe disease, particularly pneumonia in the elderly, but even so, they have not been given official names—to this day they are referred as 229E, NL63, OC43, and HKU1.

Coronaviruses are so called because the surface spikes on the virus particles give them a crown-like appearance. They are an ancient virus family, their common ancestor dating back to at least 8,000 BC. These viruses are very widespread in nature, with reservoirs mainly in bats and birds, very mobile mammals that can carry the viruses huge distances. Spillover from these virus carriers causes regular epidemics among livestock, only a few of which can be prevented by vaccination. The human corona-viruses probably all initially jumped from bats, sometimes via an intermediate host. For example, the genome sequence of coronavirus OC43 identifies its closest relative as bovine corona-virus, and its evolutionary tree reveals the date for its first spill-over to humans as around 1890. Interestingly, this date coincides with the outbreak of the flu-like illness dubbed 'Russian flu' or 'Asiatic flu', originating in Central Asia and first reported in 1889. This virus subsequently went global, causing a pandemic of a severe respiratory disease, particularly fatal in the elderly. By the end of 1890, it had killed an estimated 1 million people out of a population of 1.5 billion—among the deadliest pandemics of all time.

Unsurprisingly, there is much speculation that these two events are linked, that is that Russian flu was not flu at all but caused by spillover of coronavirus OC43 from domestic cattle. Unfortunately, as there is no stored biological material from this pandemic, it is impossible to prove the theory. Nevertheless, it is not just the coincidence of dates that is intriguing. Records clearly state that the disease specifically targeted the elderly, and this is the hallmark of the SARS and MERS coronaviruses, but not of flu (see Table 3). So assuming that the so-called Russian flu pandemic was in fact the first OC43 pandemic, it provides an interesting evolutionary famework on which to base our predictions for the long-term trajectory of COVID-19.

The future of COVID-19

As discussed earlier in the chapter, left to its own devices, COVID-19 would travel in waves around the world for many years before herd immunity could be achieved. And, as the current slogan 'no one is safe until all are safe' suggests, in our interconnected world, this means every last outpost. Yet, as herd immunity increases so the virus comes under more and more pressure to find susceptible hosts, and then virus variants come to the fore. We have already seen variants arise that spread more easily between hosts; undoubtedly these will increase in number. But, just like the evolution of the milder coronaviruses, alongside these variants, competition for hosts, most of whom are immune, should allow variants that cause less severe disease to predominate, as killing the host becomes counterproductive. Eventually these milder variants will take over, so, perhaps in 100–200 years, we will have another human common cold coronavirus to add to the list.

Past acute emerging virus infections described in this chapter amply demonstrate the enormous advances that have been made in the fight against viruses. Since the second half of the 1900s, we have moved from the familiar cyclical epidemics of acute virus infections which killed and maimed millions, to their prevention by vaccination. This has been of great benefit worldwide. We have explored how the evolutions of past virus infections are relevant to the COVID-19 pandemic in predicting how this might play out in the longer term. But not all viruses can be defeated by vaccines, and in the next chapter we discuss some of the more intransigent and persistent virus infections.

6

LIFELONG RESIDENTS

Persistent viruses, like the herpes and papilloma viruses, hide in your body for life after primary infection, and so have a very different lifestyle from those causing epidemics and pandemics of acute infections like COVID-19 and measles. As we saw in Chapters 3 and 4, viruses causing acute infections live nomadic lives and are constantly on the move in search of a temporary home—that is, a susceptible host. To maximize their chances of jumping to a new host, these viruses must inhabit crowded places, and have to produce as many new viruses as possible during the brief window of opportunity between infection of an individual and elimination by their immune response. In contrast, persistent viruses put their efforts into evading the host's immune response so that they can lodge there for a lifetime. Since they do not have to move rapidly from one susceptible host to another, they can survive where host populations are sparse and contact between isolated groups is limited. These viruses have even been found in such remote and isolated communities as the indigenous people of the South American rain forests and the highland tribes of Papua New Guinea, before these people had contact with the outside world.

Many persistent viruses are ancient human parasites; they survived in our hunter-gatherer ancestors, and were even around well before *Homo sapiens* evolved. In fact, they co-evolved with our early primate ancestors to establish a survival strategy which now only rarely becomes unbalanced and leads to significant disease. Hence one of the features of these ancient virus families which contributes to their success is that under normal circumstances infections are not life-threatening. Clearly, it would be counterproductive to kill off their long-term hosts, so they tend to cause a mild illness, or no disease at all, when they first infect, slipping into their host and taking up residence virtually unnoticed.

The herpesvirus family

Members of this large virus family are among the most common viruses to infect humans, and are also very widespread throughout the animal kingdom. They are large, DNA viruses which all establish a persistent infection in their hosts. There are eight human herpesviruses, including varicella zoster virus (VZV) and herpes simplex viruses (HSVs) 1 and 2. Both these viruses are extremely common, infecting more than 90 per cent of the world's population.

VZV spreads by close contact and through the air, and when it first infects it causes chickenpox (also called varicella), a generally mild, acute, childhood infection with a typical, body-wide rash. Once you've had chickenpox you are then immune and will not get it again. This sounds exactly like one of the acute childhood infections described in Chapter 5, but there is a significant difference. On recovery the virus is not eliminated from the body, but stays hidden in nerve cells for life in a so-called latent state. Obviously if that was the end of the story then it would be the end of the virus too, because it would die with its host. So occasionally, often several decades after the original chickenpox, the latent virus reactivates in the nerve cell it's lodging in. It travels down the nerve fibre to infect the area of skin which that particular nerve supplies. This causes the localized rash of shingles, which may occur anywhere on the body, but is most common on the face. Virus hidden in cells of the trigeminal nerve produces a rash on the face. And if it travels down the ophthalmic branch of this nerve then the rash involves the eye, sometimes with serious consequences. Nerves supplying the skin of the chest or abdomen emerge from each side of the spinal cord and run round the left or right side to meet each other in the centre front—like a bear hug from behind. So, depending on which nerve is involved, shingles on the trunk forms a band around one or other side, stopping abruptly in the middle.

The words 'shingles' and 'zoster' mean 'a belt', and 'herpes' is derived from the Greek word meaning 'to creep'—presumably all describing the distribution of the rash. The Norwegians are even more explicit; their word for shingles refers to the pain and literally translates as 'a belt of roses from hell'. Whatever its name, the tiny blisters of shingles can be intensely painful, and, just like the spots of chickenpox, they are packed with viruses. So people with either type of the infection are highly infectious to those who have not had chickenpox. Yet no one catches shingles from chickenpox because the infection causing shingles comes from inside rather than outside the body (Figure 27).

HSV, like VZV, leads a double life. In one state, it is completely hidden, latent inside a nerve cell, and in the other it produces thousands of new viruses that can spread to new hosts. This very successful lifestyle is achieved by infecting the same two cell types as VZV—skin and nerve cells. There are two types of

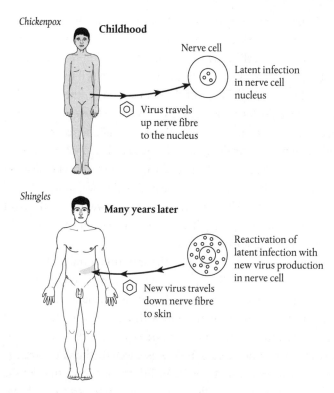

Figure 27. Latency and reactivation of varicella zoster virus. During chickenpox, the virus travels from infected skin up nerve fibres to the nerve cell nucleus where it remains in a latent form for life. In reactivation, the virus travels down the nerve again to cause the rash of shingles.

HSV—1 and 2, both spread by close contact. But while primary infection with HSV-1 causes a cold sore on the face, HSV-2 is spread by sexual contact, causing genital herpes.

HSV-1 is often spread by kissing, when the virus enters the body through a tiny, generally unnoticed, break in the skin surface—usually near the lips. It then grows locally in skin cells causing the familiar, painful cold sore. The virus spreads among skin cells until the immune response puts a stop to it. But during this growth phase HSV also infects local nerve fibres. Usually, when HSV infects a cell its DNA makes straight for the nucleus, a distance of a few micrometres. But in a nerve cell the nucleus is located in the grey matter of the spinal cord, so the virus has a long way to go. It travels along the nerve fibre towards the central nervous system at a rate of about 10 millimetres per hour, although when it finally gets to the nucleus it cannot go through its normal virus reproductive cycle because the chemicals it requires for the process are absent. But this was not a wasted journey. Like a seed lying dormant in the soil waiting for the right conditions to bring it to life, the virus lodges in the nerve cell nucleus in a latent form.

Meanwhile, back at the skin, the infection attracts the attention of immune T cells that gather to eliminate it and the cold sore heals up and is forgotten. Antibodies made in response to the infection prevent reinfection with HSV-1 in the future, but neither immune T cells nor antibodies can rid the body of the virus hidden in nerve cells. HSV-2 uses exactly the same strategy as HSV-1 to establish a persistent infection, except that HSV-2 infects and grows in cells of the genital tract and establishes latency in pelvic nerve cells.

For latent infection to succeed viruses must hide in very long-lived cells. Nerve cells are ideal for this because they generally do not divide, age, or die—those we are born with have to suffice for the whole of our lives. So there is little chance of the virus getting lost, as it might if it hid in a rapidly dividing cell with a short life span.

Yet, long-term survival of the virus depends on it completing its life cycle and producing virus offspring that emerge from the body to infect new hosts. So viruses have periodic bursts of activity, called reactivations, when thousands of new viruses are produced. The months or years of silence between these episodes maximize the chances of the virus finding susceptible hosts to infect—a recent friend or partner, or perhaps a new baby in the family. This strategy is so successful that HSV and VZV lodge in almost everyone. Just under half of all adults regularly suffer cold sores, and the virus typically passes to a young child through a kiss. Similarly, most people have shingles at least once during their adult life, when VZV can jump to susceptible young children and perhaps trigger a chickenpox outbreak.

Exactly how HSV and VZV reactivate from a latent state is not known, although slight tipping of the balance between viral latency and host immunity must be involved. Certain triggers for reactivation are common, and many people know what reactivates their own HSV infection to cause a cold sore—be it fever, stress, a cold, or menstruation. Ultraviolet light is also a frequent trigger, and some unfortunate people get a cold sore every time they go on a skiing holiday. Whatever the trigger, somehow the virus is suddenly able to reactivate in the nerve cell in which it has been latent for so long. New viruses travel down the nerve fibre, infect the skin, and a new cold sore (HSV-1) or shingles rash (VZV) appears. The same process is repeated each time the virus reactivates, and with HSV, as the same nerves are involved each time, cold sores in an individual generally appear in the same place.

Despite the fact that latent viruses are supposedly invisible to the immune system, immune control must play some role in preventing reactivation because people with impaired immunity, particularly those with T-cell defects, often have major problems with cold sores and shingles. In those with severe immunosuppression, reactivated HSV-1 can cause a generalized skin infection and may even invade the brain, causing fatal encephalitis. Also, people undergoing intensive cancer treatment, for example, may suffer repeated, severe attacks of shingles, sometimes involving more than one nerve at the same time.

Unlike HSV and VZV, several other persistent viruses do not have entirely separate latent and productive life cycles. Both phases can occur in the same cell but under varying conditions. In this case there are probably always cells somewhere in the body producing virus, and as their presence attracts attention, these viruses constantly have to dodge host immunity.

Human papilloma virus (HPV) and cytomegalovirus (CMV)

These two common viruses both establish long-term infections in cells that are actively dividing but which do not age or die. These are stem cells, which, for the lifetime of the host, constantly replenish cells of a particular lineage or type that are lost through ageing or wear and tear. The DNA genomes of CMV and HPV bed down in the nuclei of stem cells and are replicated along with the cell's own DNA. When a virus-carrying stem cell divides into two daughter cells both will contain the viral DNA, but while one of these cells remains a stem cell, the other is destined to grow into a mature, functional cell of its lineage, and eventually die.

Certain strains of HPV cause warts on the hands and verrucas (plantar warts) on the feet by infecting and establishing a latent infection in the stem cells of the innermost layer of the skin that divide constantly to replenish those lost from the surface. Maturing skin cells work their way up the multiple layers of skin to the surface, where they die and form part of the insensitive outer surface which is constantly being shed. Meanwhile, HPV follows along. As the cells mature, their internal composition changes so that at some point they can provide the necessary chemicals for HPV to reactivate and produce new viruses. So by the time the mature infected cells reach the skin surface and are about to be shed, they contains thousands of viruses ready to spread to new hosts.

Cytomegalovirus (CMV) infects around 50 per cent of adults in westernized societies, although this figure is much higher in developing countries. CMV usually infects unnoticed early in life. It establishes a persistent infection during which the virus can be found in blood, saliva, urine, genital secretions, and breast milk, and is spread through contact with these body fluids.

Once inside the body, CMV seeks out those bone marrow stem cells that will eventually mature into macrophages—a type of blood cell which patrols the tissues looking for, and removing, any foreign or dead material. So CMV is transported all over the body inside these cells, and in certain tissue sites, particularly the lungs, the virus can reactivate to produce new viruses. During this process, viral proteins appear on the surface of the infected cells, and in people with healthy immune systems these are recognized by immune T cells which eliminate most virus-producing cells before the new viruses mature. Nevertheless, some must escape immune destruction to account for the virus in body fluids. Generally no disease ensues, but in people with immune deficiencies, CMV can spread in the tissues and cause severe, even fatal, disease.

CMV infection of the lungs—pneumonitis—is a constant problem in recipients of bone marrow or organ transplants, because they have to take immunosuppressive drugs to prevent their organ being rejected. CMV pneumonitis is especially common in transplant recipients who are non-immune before the transplant but who get an organ from a CMV-infected donor.

In this situation the virus is very likely to be transferred along with the donor organ. Then, primary CMV infection occurs immediately after transplant, when the dose of immunosuppressive drugs is high. This situation can be life threatening but it can be avoided if matching donor and recipient for CMV status is possible.

CMV infection during pregnancy can also cause problems. The virus may cross the placenta and infect the developing baby. If this occurs in early pregnancy then it can cause severe abnormalities of the developing brain, lungs, and liver—but fortunately this so-called cytomegalo-inclusion disease is very rare.

Immune evasion

For persistent viruses, evading immune recognition is key to their survival, and they show astonishing ingenuity in establishing and maintaining a hold on their host. Even if they go into hiding inside a cell and are completely indistinguishable from the host, from time to time they have to reactivate. So these viruses have evolved a whole gamut of mechanisms designed to extend their reproductive phase for as long as possible by subverting or suppressing the immune response directed against them. Since T cells are the most effective weapon the body has against viruses, it is here that a virus's efforts are concentrated.

All mammals have immune T cells that, in theory, are capable of killing indiscriminately—even attacking their own body's ('self-') cells. So T cells have evolved to distinguish between host proteins and foreign, or in this case virus, proteins. Cells have a complex series of reactions that break proteins up into short pieces called peptides and display these on the cell surface. Passing T cells interact with these molecules, and if they recognize them as 'self-peptides' then the cell is left intact. But in a virus-infected cell, viral peptides are also displayed on the cell surface, and these act as flags to alert T cells to kill the cell. As a countermeasure, persistent viruses have evolved many different mechanisms to camouflage their presence, which researchers are still unravelling. Suffice it to say that the pathway that converts a cellular or viral protein into a peptide on the cell surface involves many different molecules with individual functions. Viruses strike at many key points along the way to interrupt the whole process and, interestingly, several completely unrelated viruses have evolved very similar mechanisms for subverting this pathway—an indication of how important immune evasion is for their survival.

Although each virus appears to have an infallible counter-attack to the onslaught of the immune system, coevolution has ensured that most persistent viruses establish a balance between successful virus production and immune recognition. The balance may be set at slightly different levels for each virus, so that some are easier to find in the body than others, but usually the virus infection exists at a level which does not inconvenience the host. The exception to this is HIV which, probably because it has only recently infected humans, has not yet established this lifelong balance. We discussed the emergence of HIV in Chapter 4; here we consider how the virus persists in the body and causes AIDS.

Retroviruses

Retroviruses are so called because of their unique survival strategy. They perform the trick of reverse transcription as soon as they infect a host cell, and once integrated, their genome can remain there for the lifetime of the cell. If the cell divides, the viral DNA is automatically reproduced along with it. In this way, there is no need for the virus to express any of its own proteins, so the infected cell does not invite immune attack.

Human immunodeficiency virus-1 (HIV-1)

HIV-1 was first isolated from the lymph gland of an AIDS sufferer by Francoise Barré-Sinoussi at the Pasteur Institute in Paris, in 1983.[1] She and the head of the research group, Luc Montagnier, won the Nobel Prize in Physiology or Medicine for this discovery in 2008. But almost as soon as HIV-1 was first described in the scientific literature it invited controversy. Montagnier innocently sent samples of the virus to a well-known American retrovirologist, Robert Gallo at the National Institute of Health (NIH) in Bethesda, Maryland, US, for collaborative studies. Then in 1984, Gallo's research group published a scientific article apparently describing their own HIV-1 isolate, which they succeeded in growing in the laboratory in sufficient quantity to develop a lucrative blood test for HIV-1.[2] But when both the French and American groups published the genetic sequences of their respective HIV-1s it was obvious that they were in fact the same virus isolate. The implication was that Gallo had in effect 'stolen' the French virus sent to him in 1983, presenting it as a new isolate in his 1984 article. There followed a wrangle over the rights to the diagnostic blood test which attracted worldwide attention. So much was at stake, both in terms of finance and prestige, that the dispute went to the very top. It was eventually settled when American President Ronald Reagan and French Prime Minister Jacques Chirac agreed to split the credit between the two scientists and the royalties between the two countries.

All returned to normal after this debacle, but not for long. A small but vociferous band of scientists, led by American retrovirologist Peter Duesberg, believed that HIV-1 did not cause AIDS. He maintained that scientists wished to convict an innocent virus as the cause of AIDS, saying that the association was an invention by desperate virologists looking for a reason for their existence. He felt so strongly about this that, in 1996, he wrote a 700-page book called *Inventing the AIDS Virus.*[3]

In their heyday Duesberg's maverick views were championed by some in the media, and not only had dangerous implications for the safe sex campaign but also promoted the 'denial theory'. HIV-positive people and their families, while struggling to accept the awful truth, were only too willing to believe propaganda telling them that the virus was harmless. This could have induced those infected but feeling perfectly well to abandon their treatment and even stop taking the precautions necessary to prevent passing the deadly virus on to others. Fortunately, by the mid 1990s, the success of antiretroviral therapy at last caused Duesberg to lose his constituency.

The natural history of HIV-1 infection outlined below prevailed for several years until the discovery of life-saving antiretroviral therapy. In affluent countries these drugs eventually transformed the infection from a death sentence into a manageable disease allowing a normal lifespan. But this is not the case elsewhere. Despite the efforts of many governments and charities, by 2019 UNAIDS calculated that around 32 per cent of HIV-infected adults and 47 per cent of infected children worldwide still had no access to these essential treatments, and so the death rate remains extremely high.

HIV-1 spreads from one person to another via contact with body fluids, including blood, breast milk, semen, and vaginal secretions, all of which contain both virus particles and virus-infected cells. HIV-1 uses the cellular CD4 molecule as its receptor for attaching to and entering cells. Thus the distribution of

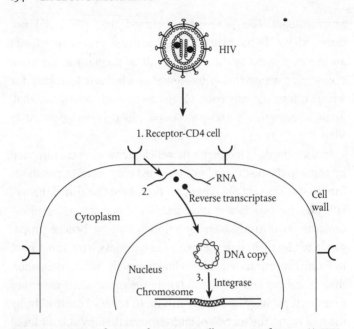

Figure 28. HIV-1 infection of a CD4 T cell. During infection: (1) HIV-1 attaches to CD4 on the cell surface; (2) virus RNA is released into the cytoplasm where the reverse transcriptase enzyme converts it into DNA; (3) the viral enzyme integrase integrates the viral DNA into the cell's chromosomes and it remains there for the lifetime of the cell.

this molecule in the body, mainly on helper T cells, but also macrophages and microglial brain cells, determines the profile of the disease (Figure 28).

When HIV-1 first infects, most people suffer a mild flu-like illness, but occasionally a more severe disease ensues, with fever and swollen glands resembling glandular fever (or infectious mononucleosis). At this stage, levels of HIV-1 in the blood are very high, as the virus inside CD4 cells spreads throughout the body and lodges in the brain. This infection is brought under control two to three weeks later, when the immune response is fully operational, and the person recovers to remain healthy for a decade or so. But the virus cannot be eliminated from the body because it is integrated into cellular DNA and is reproduced each time an infected cell divides.

During the early stages of the asymptomatic period, levels of HIV-1 in blood are low, as any infected CD4 cells expressing HIV-1 proteins will invite immune attack and antibodies and killer T cells control virus spread in the body. But HIV-1 is adept at evading immune mechanisms, mainly by mutation. Each time the virus reproduces it throws up mutants, some of which will be unrecognizable to killer T cells. These mutants then have a selective advantage over their non-mutated counterparts, and they take over—that is, until the immune system catches up with them. But new mutants keep arising, and the process is repeated over and over again. Because CD4 T cells that the virus infects and destroys are pivotal to the immune response by kick-starting B cells to make antibodies and T cells to kill virus-infected cells, the body is caught in a downward spiral. Ultimately, the supply of CD4 cells runs dry, and with the resulting immuno-deficiency mutant viruses run riot. This is the beginning of clinical AIDS, when opportunistic infections eventually kill the patient (Figure 29). Virus-associated cancers, called 'opportunistic neoplasms', are also more common in AIDS patients, and these are discussed in Chapter 7.

Hepatitis viruses

Several viruses, including hepatitis A and E viruses, cause acute inflammation of the liver, or hepatitis, but only two can persist in the body for life after primary infection. These are hepatitis B virus (HBV) discovered in 1964, and hepatitis C virus (HCV)

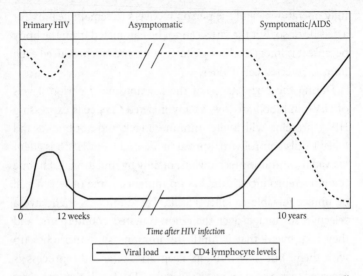

Figure 29. Time course of non-treated HIV infection showing CD4 cell count and viral load during primary, asymptomatic, and symptomatic phases of HIV infection.

discovered in 1988. Prior to their isolation these viruses together constituted a major hazard for recipients of transfusions of blood and blood products.

Modern blood banks, with highly organized systems for collection, processing, testing, and storage of donor blood, came into existence during the Second World War because of the very high demand for blood transfusions. At the time a blood transfusion was a hazardous procedure, not least because of the unknown, blood-borne viruses that were infused along with the life-saving blood. It was not long before post-transfusion hepatitis was reported, the hallmark being jaundice, sometimes accompanied by nausea, vomiting, malaise, and fever, occurring

1–4 months after a transfusion. Studies in the US estimated the incidence of carriers of a putative hepatitis virus to be 6.3 per cent among commercial (paid) donors but less than 0.6 per cent for volunteer donors.[4] An analysis of patients undergoing open-heart surgery, who require many units of fresh blood, revealed an incidence of post-transfusion hepatitis of a massive 51 per cent. A viral cause was suspected, and the race was on to find the culprit. This was the backdrop against which Baruch Blumberg, working at the Fox Chase Cancer Center in Philadelphia, happened upon that very agent. But the discovery was so serendipitous that it took several years before the truth was uncovered, generally accepted, and acted upon.

Hepatitis B virus

Blumberg's research topic was 'inherited and environmental variation in disease susceptibility', an interest which he developed as a medical student in New York when he became 'aware of differences among individuals and among populations . . . in responses to disease-causing agents'.[5] This led him to travel the world collecting blood serum samples from diverse populations. Returning home, he and his team then tested these samples against serum from people who had received multiple transfusions, and who, Blumberg reasoned, might have developed antibodies against foreign serum proteins in the infused blood. The hope was to uncover congenital differences, or polymorphisms, that would account for variable disease susceptibility. A reasonably logical approach, but a bit like looking for a needle in a haystack. However, they found exactly what they were looking for!

In 1964, while testing sera from Australians, including some First Nation Australians, against serum from a multi-transfused haemophiliac patient, Blumberg's team identified a reaction which indicated that the haemophiliac had antibodies against something in the serum of one of the Australian blood samples. On further analysis this 'something' turned out to be absent from sera from healthy US citizens but common among people from Taiwan, Vietnam, Korea, and the central Pacific, and in First Nation Australians. This gave the team a new protein to work on, which they called the 'Australia antigen' (Au).

Over the next year the group continued to plot the demographics of Au positivity and antibodies against it. South and Eastern European countries were added to the list of countries where the antigen was common, although antibodies against it remained most common in multi-transfused people. Slowly, the realization that Au could be a transfused agent began to dawn, but it still took several more pointers before the eureka moment arrived. One such pointer was a longitudinal study performed on an institutionalized patient from New Jersey who had several negative tests for Au before a surprising positive one. This coincided with a diagnosis of clinical hepatitis, and so the team were stimulated to test for Au in other hepatitis patients. They showed significantly more Au positives in the hepatitis than the control group, but their scientific paper reporting these results was rejected for publication on the basis that it was a 'false claim', coming from a group with no experience in virology or transfusion medicine. The group did eventually publish their findings in 1967, and from then on recognition of their work slowly grew.[6] Au turned out to be the hepatitis B surface antigen, and when screening of blood donations for this protein began in 1971, blood transfusion became a whole lot safer. Blumberg and his team won the Nobel Prize for Physiology or Medicine for their discovery in 1976.

HBV is a highly infectious DNA virus that is spread by close contact with a carrier, including sexual contact and mother to child, as well as blood contamination of medical and dental equipment, shared household utensils, tattooing, and body piercing. Intravenous drug users and gay men are particularly at risk.

While just 1–5 per cent of HBV infections in adults progress to a persistent infection, virus passed from mother to child at birth establishes lifelong infection in over 90 per cent of untreated cases. Because the liver is a large organ with plenty of spare capacity, persistent HBV infection often passes unnoticed for several years despite cumulative liver cell damage that may lead to cirrhosis and liver failure. Worldwide, the WHO reports around 257 million people infected with HBV, causing approximately 887,000 deaths annually, the highest prevalence being in Southeast Asia and sub-Saharan Africa. However, vaccines against HBV have been available since 1982, and have gone a long way to protecting those at risk of infection. In particular, vaccination immediately after birth can prevent mother-to-child transmission, and almost certainly virus persistence, in the newborn.

As it turned out, the introduction of HBV donor screening did not entirely remove the risk of post-transfusion hepatitis. Up to 10 per cent of recipients still developed the disease; an indication that another, as yet unknown virus, was at large. This was called non-A, non-B hepatitis virus, a name that vanished after hepatitis C virus was isolated.

Hepatitis C virus

In the 1970s and '80s there were many false trails to the discovery of HCV, but eventually in 1988, Michael Houghton and his team at Chiron Corporation, Emeryville, California, succeeded.

They used molecular techniques to isolate short pieces of the viral genome sequence from the plasma of a chimpanzee that had been infected with the non-A, non-B hepatitis virus through inoculation of plasma from a patient with post-transfusion hepatitis. The entire genome of this small RNA virus was sequenced within a year and donor screening was implemented immediately, even before the virus was visualized under the electron microscope. With several refinements to the testing regime over the years, the risk of HCV infection has now dropped to 1 in 100,000 per unit of infused blood. In 2020, Houghton and colleagues were awarded the Nobel Prize in Physiology or Medicine for their work on HCV.

HCV is mainly spread by blood-contaminated injecting equipment, and is therefore common among intravenous drug users. As primary infection is usually asymptomatic, most carriers are unaware that they are infected with a killer virus. Nevertheless, the virus homes in on the liver where it establishes a lifelong infection. Replication in the liver over years causes liver damage predisposing to cirrhosis and liver failure. The WHO estimates that 71 million people globally carry HCV, which accounts for nearly 400,000 deaths annually.

Several persistent virus infections, including both HBV and HCV, are linked to cancer development, and this association is discussed in Chapter 7. In addition, the WHO has set the goal of eliminating viral hepatitis as a public health threat by 2030.

Virus links to orphan diseases

Viruses are often used as scapegoats, particularly for chronic diseases with no known cause, and have been suggested as the culprit in diseases as diverse as multiple sclerosis, heart disease, and diabetes. In each case the hunt for a specific virus has proved fruitless. The crucial question is, could a common, well-known virus occasionally change its ways and cause a quite different disease pattern? There are several proven examples of this, including paralytic polio discussed in Chapter 5; three more examples are outlined below:

- First is the herpesvirus, Epstein–Barr virus (EBV). This virus usually silently infects young children and thereafter remains in the body for life, carried by B cells in the blood. As we have seen, this pattern of silent, childhood infection is typical of herpesviruses, and EBV infection is almost universal in developing countries, where over 95 per cent of children are infected by three years of age. The virus is present in saliva and spreads by close contact, such as kissing or sharing utensils. But children in countries with high standards of hygiene may miss out on infection early in life. They are then susceptible to infection as teenagers and young adults, when kissing is the usual mode of transmission. This delayed infection, rather than being silent, is often manifest as glandular fever (infectious mononucleosis)—aptly named the 'kissing disease'. Common among students, the disease is characterized by fever, sore throat, enlarged lymph glands, and fatigue that often persists for several months before full recovery. We will meet EBV again in Chapter 7 to discuss its cancer-causing potential.
- Second is measles virus, which can also change its spots from its usual pattern of childhood infection to something much more sinister. Measles is an acute illness with fever and a rash which lasts 1–2 weeks before the virus is eliminated from the body never to infect again. But very rarely,

usually when it infects a very young child, measles virus persists in the brain, causing a degenerative disease known as subacute sclerosing panencephalitis. This manifests up to ten years after the initial infection and is invariably fatal. But fortunately, as vaccination has eliminated the virus from many parts of the world, this complication of primary infection is also vanishing.

- Third is persistent HSV that occasionally causes herpes stromal keratitis, a disease of the cornea of the eye that can lead to blindness. This is thought to be an autoimmune disease, a group of diseases with a potential viral cause. Viruses normally trigger killer T cells directed against cells expressing viral peptides. But occasionally, quite by chance, the viral peptide that these T cells recognize is identical to a normal cellular peptide. So T cells generated to fight the virus mistakenly recognize and kill cells expressing this particular peptide as well. This is called 'molecular mimicry' and it generates an autoimmune reaction—so called because the immune response (antibodies and/or killer T cells) is misguidedly directed against self (auto) (Figure 30). In this case the virus acts as the trigger to set off a reaction which continues long after the virus has gone. It is postulated that because of molecular mimicry between HSV and corneal cells, T cells recognizing a peptide from a viral coat protein also target and destroy the cornea.[7]

But just because herpes stromal keratitis is thought to be caused by a virus does not mean that all autoimmune diseases follow the same pattern. Multiple sclerosis (MS) is a case in point. This terrible disease apparently strikes out of the blue, favours young adults, and can cause devastating and eventually fatal damage to the central nervous system. The disease usually progresses spasmodically, with relapses and partial recovery, so that permanent damage to the nervous system accumulates over time. Relapses often seem to follow an acute viral infection, but although virologists have undertaken exhaustive searches, no consistent virus association can be found. But still many scientists believe that MS must be triggered by a virus.

Over the years many viruses, including measles, EBV, canine distemper virus, and human herpes viruses six and seven, have all been heralded as the cause of MS but then dropped through lack of evidence. Yet recently the case for EBV has become more compelling, but is not proven. The features of MS fit with both the EBV and herpes simplex models outlined above. For one thing, the demographics of MS resemble those of glandular fever. Both diseases target young adults in affluent Western-style societies; both are rare in non-industrialized, tropical countries. The geographical variation seen with MS led to a study of people who have moved from high- to low-incidence areas and vice versa. It turns out that if young children move, then they adopt the risk of MS found in the new area; but when adults move, they retain the risk of their original home. This suggests that, whatever the predisposing factor for MS is in the high-incidence areas, it is acquired during childhood and then retained for life. Is it a common childhood infectious agent, perhaps a virus like EBV or measles? This putative virus would infect most people harmlessly as children, but in a few cases, possibly as a result of delayed infection, would cause or predispose to MS in later life.

The brains of MS sufferers contain large numbers of immune cells that attack normal nerve cells—particularly targeting a protein called myelin basic protein. This strongly suggests that MS is an autoimmune disease with molecular mimicry between an unidentified viral protein and myelin basic protein being the

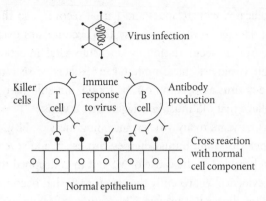

Virus infection

Killer cells | T cell | Immune response to virus | B cell | Antibody production

Cross reaction with normal cell component

Normal epithelium

Figure 30. Molecular mimicry. Virus infection induces antiviral T- and B-cell responses. Killer T cells and antibodies then recognize and target a normal cell component as well as a viral component, causing auto-immune disease.

cause of the damage. A causative link to EBV is based on the fact that virtually all MS sufferers are EBV positive, but then so are the vast majority of healthy adults—the difference being 100 per cent positive in MS versus 90–95 per cent in healthy adults in high-risk MS areas. Overall, EBV carriers are 15 times more likely to develop MS than matched, EBV-negative controls, and the lifetime risk of developing MS is increased two- to three-fold in people who have suffered from glandular fever. So could T cells directed against EBV proteins occasionally recognize neural proteins? Only time, and more detailed research, will tell.

To summarize, persistent viruses tend to strike up stable relationships with their respective hosts as they skilfully dodge immune destruction and exploit the host to ensure their own long-term survival. This is an incredibly successful lifestyle for a virus, which generally causes little harm to the host. But still there can be problems. The most obvious of these is seen with immunosuppression of the host leading to virus reactivation and disease, but there are also more subtle, long-term effects.

A cell that is persistently infected by a virus carries a set of foreign genes which are passed to all its daughter cells. Although it may not produce overt disease, this lifelong intimacy between the virus and the cell can influence the destiny of both parties. In some cases persistent virus infections eventually lead to cancer, and this topic is the subject of the next chapter.

7

VIRUSES AND CANCER

Cancer is a common and deadly disease which comes in many different forms. Overall, around half of the population suffers from cancer at some time in their lives, and, despite all the recent advances in scientific techniques and treatment and prevention strategies, approximately half of those diagnosed with cancer die of it within ten years.

A cancer arises from a single cell in the body multiplying unchecked until it has produced a whole mass of identical cells—a tumour. This can happen in any organ of the body, in people of any age, and in any country; no one is exempt. Paradoxically, although it is common to meet people with cancer, since each person is made up of around 10^{14} cells, at the level of individual cells cancer is extremely rare. So what causes a cell to turn cancerous?

A cancer is the final result of damage inflicted on a cell which may be caused by a multitude of different factors. Most cancers seem to appear for no reason in perfectly healthy people and the cause is often unknown. But the development of a cancer may be linked to a person's habits, lifestyle, or occupation. That smoking tobacco predisposes to lung cancer and excessive sunlight can cause skin cancer is common knowledge, yet there are many other cause-and-effect associations. Historically, cabinet-makers, who inhaled the fine dust generated by sawing hardwoods, had an increased risk of cancer of the nasal sinuses; boy chimney sweeps, employed to climb up inside chimneys, often suffered from cancer of the scrotum caused by fine soot powder getting lodged in skin creases; and people working with asbestos risked developing a mesothelioma (a cancer of the lining of the lung). More recently, the knowledge that exposure to radiation increases the incidence of almost all forms of cancer came from the survivors of the atomic bombing of Nagasaki and Hiroshima in 1945. Similarly, in the aftermath of the 1986 Chernobyl nuclear power plant disaster, a distressingly high number of children developed cancer—particularly involving the thyroid gland.

These examples link cause and effect, but tell us little about the all-important changes required in a single cell to transform it from a normal to a cancerous one. The details of these cellular hits slowly became apparent after the discovery of DNA in the 1950s and the molecular revolution beginning in the 1960s that, in effect, allowed scientists to dissect cells and unravel the function of single genes in their growth control mechanisms. A few direct cancer-causing genes were discovered along this journey, including the BRCA 1 and 2 genes that increase the chances of breast and ovarian cancer. But on the whole, as we will see, the journey towards cancer is much more tortuous.

Genes that induce cells to divide, called oncogenes, were first discovered in the retroviruses that cause tumours in chickens, and were only later found to arise from cellular genes that had been hijacked by these viruses. The function of oncogenes in driving cells to divide is balanced by tumour-suppressor genes, also called anti-oncogenes, that stop cells dividing. The existence of these genes was postulated in 1969 by cell biologist Henry Harris at the University of Oxford, for, as he described it, 'putting the brakes on' cell division,[1] and, sure enough, they were discovered in 1971. In 1976, J. Michael Bishop and Harold E. Varmus at the University of California, San Francisco, discovered the first oncogene in the human genome, for which they received the Nobel Prize in Physiology or Medicine 1989.[2] We will come back to these genes that control cell division later.

We now know that worldwide around 15 per cent of cancers are caused by viruses, and this figure rises to 50 per cent in low- and middle-income countries. In this chapter we look at the history of tumour virus discovery and at how the dramatic effects of these viruses are intimately linked with the genes that control normal cell growth.

Long before viruses were actually seen under the electron microscope, scientists were hunting for filterable agents that were associated with cancers in animals. In 1908, two Danish scientists, Wilhelm Ellermann and Oluf Bang, carried out experiments which suggested that leukaemia in chickens was infectious. After making an extract of leukaemia cells and passing it through filters which trapped even the smallest bacteria, they found that this filtrate could transmit leukaemia to healthy chickens. But back then leukaemia was not regarded as a malignant disease, and so the discovery passed virtually unnoticed.

Then, in 1911, Peyton Rous at the Rockefeller Institute in New York, who was also working with chickens, carried out a similar experiment using an extract made from a solid tumour. This tumour came from the right breast of a barred Plymouth Rock hen brought to him by a Long Island chicken farmer. The hapless farmer hoped that Rous could cure the hen, but Rous promptly killed it off and used the tumour for his own research.

He injected a tumour extract into healthy chickens and produced the same type of tumour.

For a long time the importance of this work was not accepted by other scientists, particularly those engaged in cancer research. They felt sure that toxic chemicals were the cause of all cancers, and simply could not believe that cancer could be caused by a virus because it was clearly not infectious in the same way diseases like measles and chickenpox were. So Rous had a hard time gaining recognition for his work against the closed minds of his contemporaries, and eventually turned to experimenting with toxic chemicals himself. But slowly more evidence accumulated and, more than 50 years later, Rous was awarded the 1966 Nobel Prize in Physiology or Medicine for his groundbreaking work on the so-called Rous sarcoma virus.

The next breakthrough came in 1936 from John Bittner, who was studying breast cancer in mice at the Jackson Laboratory in Maine, US. He used two strains of mice—one with a high incidence of breast cancer and the other with a low incidence—to carry out a simple but convincing experiment. In a classic swap-over design, he took newborn mice from the two strains away from their natural mothers and put the offspring of high-tumour-incidence mothers to be suckled by low-incidence foster mothers and vice versa. Then he just waited to see if the incidence of cancer in the fostered mice differed from the control mice left with their natural mothers. This simple experiment showed conclusively that the high tumour incidence was caused by an agent transmitted in breast milk from the high-incidence foster mother and not, as was generally believed at the time, by an inherited genetic susceptibility from the biological mother. That agent in milk turned out to be a virus, and with this discovery the scientific community finally sat up and took notice—the era of tumour virology had begun.

But after the successes in animal tumour virus research, years of disappointment lay ahead for human tumour virologists. Although there are many human cancers, both solid tumours and leukaemias, that resemble cancers caused by viruses in chickens and mice, no human tumour viruses were discovered. Many scientists began to doubt whether any human cancers were caused by viruses, while others turned to epidemiologists for help in uncovering clues at the population level. They reasoned that if a tumour is caused by a virus then it should be infectious and spread from one person to another by one of the routes commonly used by other viruses. So, depending on its method of spread, the tumour might, for instance, show geographical variation like arthropod-transmitted viruses, or be more common among family members, close associates, or people with the same occupations or habits than in the general population. Also, people who are highly prone to virus infections of all kinds because of immune suppression might be more susceptible to tumours caused by viruses. These were the clues that tumour virologists were looking for in the early 1960s when the following event took place, which proved a milestone in the history of tumour virology.

The discovery of Epstein–Barr virus (EBV)

The virologist Anthony Epstein was working on Rous sarcoma virus at the Middlesex Hospital in London when, in 1961, he attended a lecture entitled: 'The commonest children's cancer in tropical Africa—a hitherto unrecognised syndrome'.

The lecture was given by Denis Burkitt, a missionary surgeon working in Uganda. He described tumours of the jaw in African children which grew rapidly and killed in a matter of months.

To the audience this was a new type of tumour not seen in Britain, Europe, or the US, and therefore fascinating in itself. But even more intriguing was Burkitt's description of the geographical variation in tumour incidence; in some Ugandan villages there were many children with tumours while in others there were none.

In 1961, Burkitt set off to investigate further. Travelling in a second-hand station wagon on what he called his 'long safari'—a 10,000-mile tour of East Africa—he mapped the distribution of the tumour (Figure 31). Accompanied by two medical missionaries, he visited more than 50 large, small, and bush hospitals, and talked to many hundreds of doctors along the way. He always gave an amusing account of this field trip; at one stop, for example, he came across a mortuary attendant whose hobby was making clay models. He often used corpses from the mortuary as models for busts—presumably because they remained still for longer than live subjects. Amongst his collection were

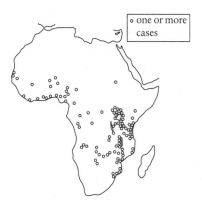

Figure 31. Burkitt's map of the distribution of Burkitt's lymphoma in Africa, plotted after his 'long safari' in 1961.

several clay heads of children with the jaw tumours that Burkitt had come to investigate. So he proudly lined them up for inspection on the concrete pathway outside the mortuary. They were so lifelike that while Burkitt was examining them someone passing by commented on the novel method of treatment—burying the patient up to the neck in concrete!

As a result of his trip, Burkitt made a detailed map of the tumour's distribution in Africa. This revealed that it was confined to low-lying areas of central Africa, that is, areas below 5,000 feet. Burkitt found that the tumour was also restricted to areas where rainfall was above 55 cm per year and where the temperature did not fall below 16°C—in other words, areas with a hot and humid tropical climate. Burkitt realized that this distribution was very similar to that of holoendemic malaria, that is, non-seasonal malaria that occurs at the same level all the year round.[3] (Later, Burkitt's tumour was also found to occur in Papua New Guinea, where the same climatic conditions prevail.)

The reason for the geographical restriction of holoendemic malaria is the life cycle of the *Anopheles* mosquito which spreads the parasite *Plasmodium falciparum* from one person to another. The larval stage requires water, hence the high rainfall requirement, and adult females do not lay eggs at temperatures below 16°C. In areas of Africa where these requirements are not fulfilled, malaria is seasonal, appearing with the rains and disappearing again in the dry season. So Burkitt put two and two together and suggested that the jaw tumour was caused by an infectious agent that was spread by a mosquito vector. In fact, it later became clear that he was right about the infectious agent— a virus—and about the association with holoendemic malaria, but the virus is *not* spread by mosquitoes.

Listening to Burkitt's presentation, Epstein was fired with enthusiasm to hunt for a virus linked to this tumour. He immediately arranged with Burkitt for live tumour biopsy samples to be flown from Africa to his laboratory in London. But tumour cells are fragile and once removed from the body they die rapidly. So on their long journey they had to be kept alive in a special culture medium. Bacterial contamination was also a problem because if a tumour sample was inadvertently contaminated then the bacteria would grow in the culture medium at the expense of the tumour cells.

Despite all the hazards, Epstein got his samples; but disappointment was in store. For two years he and his coworkers tried to isolate a virus by all the techniques known at the time. They searched tumour material under the electron microscope for virus particles, but found none. They put small pieces of tumour into a culture of liquid growth medium closely resembling body fluids to try to persuade the cells to grow, but had no success there either. Nevertheless, Epstein persevered because, he said, 'it just had to be right'.

In 1964, when the breakthrough eventually came, it was quite by chance. One of the valuable tumour samples was held up, and finally arrived at the laboratory after many days in transit. The medium looked cloudy—normally a sure sign of bacterial contamination. But when Epstein looked at it under a microscope he found that the cloudiness was caused by floating tumour cells, shaken free from the tumour biopsy during the long journey. He put this suspension of floating cells into culture, and for the first time they grew! As soon as he had enough growing cells, Epstein examined them under the electron microscope. At last he found some virus particles. Epstein remembers, 'I simply switched the microscope off, went out and walked round the block two or three times in the snow before I dared come back and look again'.[4]

The reason that Epstein found virus particles in this particular sample, but not in the many earlier attempts, was because this was the first to grow in culture *before* being processed for electron microscopy. Culturing the cells was an essential step since, although all the tumour cells contain the viral DNA, it is in a latent form that is invisible under the microscope. When in the body, any tumour cell starting to produce virus particles immediately comes under attack from immune T cells, so only when cells were grown *outside* the body could virus particles be produced.

Epstein went on to show that the virus he discovered was a new type of herpes virus which was consistently found in these African tumours. The tumour, for obvious reasons, became known as Burkitt's lymphoma, while the virus is Epstein–Barr virus (EBV), named after Epstein and Yvonne Barr, the research assistant responsible for culturing the tumour cells.[5]

This was the beginning of a life's work for Epstein and his team. They had to find out if the virus actually caused the tumour and then convince others with their experimental results. This was not an easy job; just because a virus is found in a tumour it does not mean that it causes it. It may, for example, have infected the tumour cells after the tumour had developed. It took Epstein many years to persuade some scientists that EBV was the cause of Burkitt's lymphoma. The virus spreads by close contact, and the picture of its association with a specific type of tumour was complicated by the discovery that it is extremely common—it infects almost everyone worldwide, silently during childhood. Then in 1968, the finding that it also causes infectious mononucleosis (glandular fever) in a proportion of those whose primary infection is delayed further complicated the picture.

Since those early days EBV has been identified as the cause of several other tumours, including around 20 per cent of gastric cancers[6] as well as lymphoid tumours encompassing approximately 35 per cent of Hodgkin lymphoma,[7] rare T-cell lymphoma,[8] and B-cell lymphoma associated with immuno-deficiency.[9] In 1973 EBV was identified in the malignant epithelial cells of nasopharyngeal carcinoma,[10] a tumour of the nasal passages that is very rare in most of the world but very common in South East China, Hong Kong, and Taiwan, where the incidence reaches as high as 40 per 100,000 each year. The association between tumour and virus is virtually 100 per cent, but the reason for its clear geographical restriction is not fully understood, although it is thought to be due to a local dietary component that enhances tumour outgrowth.

Epstein's discovery of the first human tumour virus renewed virologists' flagging enthusiasm for the hunt, and opened the way for several more important discoveries. In total seven human tumour viruses are now recognized, and several show geographical restriction (Figure 32), but intriguingly, they are quite different types of virus. They include five DNA viruses, EBV and human

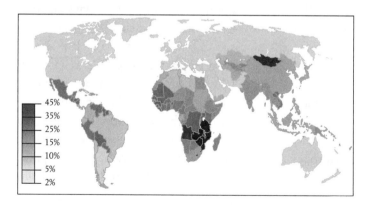

Figure 32. World map showing areas of high incidence of virus-associated tumours. Attributable fraction of cancer related to viral infections, 2012.

herpesvirus 8 (both herpesviruses), hepatitis B virus, a hepadna-virus, human papilloma virus, and, most recently, Merkel cell polyoma virus. The list is completed (for the time being) by a pair of RNA viruses—human T-cell lymphotropic virus-1 (a retro-virus) and hepatitis C virus (a flavivirus). Finally, by causing severe immunosuppression, HIV-1 enhances the oncogenic effect of all these cancer viruses, and so it is considered a cancer-causing agent by some, including the International Agency for Cancer Research (Table 4).

The six human tumour viruses in addition to EBV were identified using a variety of techniques, only one being dis-covered by cell culture techniques similar to those used by Epstein and his team—that is, human T-cell lymphotropic virus-1 (HTLV-1). Although many retroviruses cause tumours in animals—including those investigated by Ellermann and Bang, Rous, and Bittner described earlier in this chapter—HTLV-1 was the first human retrovirus to be discovered and remains the only known retrovirus to cause tumours in humans.

The virologist Robert Gallo, working at the National Institute of Health in Bethesda, US, spent many years trying to track down a human leukaemia virus by attempting to grow the malignant cells in tissue culture and then to isolate a virus from them. But, similar to Epstein's experience, this approach failed because the cells would not stay alive long enough to study them. This all changed when Gallo identified the T-cell growth factor now called interleukin-2. This stimulated the cultured leukaemia cells to grow, and Gallo then used a sensitive assay to test the cultures for production of reverse transcriptase, an enzyme produced by retroviruses. This finally rewarded him with positive results.

The first isolate of HTLV-1 came from the blood of a young black man who had an aggressive form of T-cell leukaemia.[11]

Table 4. Human Tumour viruses.

Virus	Date of discovery	Virus family/genome type	Associated tumours	Other diseases caused
Epstein–Barr virus (EBV)	1964	Herpesvirus DNA	Burkitt's lymphoma, some Hodgkin lymphoma, Nasopharyngeal carcinoma, some gastric cancer, post-transplant lymphoma	Glandular fever[1]
Hepatitis B virus (HBV)	1965	Hepadnavirus DNA	Liver cancer	Hepatitis, cirrhosis
Human T-cell lymphotropic virus-1 (HTLV-1)	1980	Retrovirus RNA	Adult T-cell leukaemia	Tropical spastic paralysis
High-risk human papilloma viruses (HPV)	1983	Papilloma virus DNA	Cancer of cervix, anal and vulval cancer, some head and neck cancers	
Hepatitis C virus (HCV)	1989	Flavivirus RNA	Liver cancer	Hepatitis, cirrhosis
Human herpesvirus 8 (HHV8)[2]	1994	Herpesvirus DNA	Kaposi sarcoma	
Merkel cell polyoma virus (MCV)	2008	Polyomavirus DNA	Merkel cell cancer	

[1] Also called infectious mononucleosis.
[2] Also called Kaposi sarcoma-associated virus.

Several more virus isolates followed, all from adults who suffered from similar rapidly fatal leukaemias. Intriguingly, all these early cases had links to either the Caribbean or Japan.

Gallo's work was published in 1980, and at the same time Yorio Hinuma and coworkers at Kyoto University in Japan published details of clusters of patients with a type of adult T-cell leukaemia or lymphoma on the islands of south-west Japan. The malignant cells from these cases all contained HTLV-1.[12]

A search for healthy HTLV-1 carriers found that overall the infection is extremely rare; the virus does not spread easily between humans and tends to infect within families. It is transmitted by blood transfusion, by sexual contact, and from mother to child via breast milk, and thereafter establishes a lifelong infection in CD4 T cells. Around one in 80 of those persistently infected with HTLV-1 eventually develop adult T-cell leukaemia, generally some 10–40 years after first infection. In the Caribbean, where the virus is relatively common, HTLV-1 also causes a chronic neurological disease called tropical spastic paraparesis.

HTLV-1 infection is virtually restricted to specific regions, but the geographical puzzle of its global distribution has still not been elucidated. Epidemiological surveys showed that between 3 and 10 per cent of healthy individuals in south-west Japan and the Caribbean were positive for the virus, and there were pockets of infection in sub-Saharan Africa, South America, the Middle East, and Melanesia. Also, retroviruses carried by African monkeys, particularly chimpanzees and African green monkeys, are 95 per cent identical to HTLV-1, and are therefore the likely direct ancestor of the human virus.

To explain these findings, Gallo suggested that HTLV-1 was carried from Africa to the Caribbean along with the slave trade during the sixteenth to nineteenth centuries. He further hypothesized that in the sixteenth century, Portuguese traders, known on occasions to travel to Japan via central Africa, took local people and animals with them, thus seeding the virus in the Japanese population. There may be truth in this theory of early virus spread, but more recent findings of HTLV-1 in further geographically and ethnically distinct populations suggest a more complicated virus-dissemination pattern.

Following the molecular revolution beginning in the 1960 and 1970s, it became more profitable to hunt for unknown viruses in human tumours using molecular probes to identify the viruses' genetic material, rather than cell culture techniques. As a young virologist, Harald zur Hausen, a molecular virologist from Germany, worked on EBV, using these new molecular techniques to identify EBV in the malignant epithelial cells of nasopharyngeal carcinoma. But when, in 1972, he was appointed to the Chair of the Institute of Clinical Virology in Erlangen-Nurnberg, he decided to change tack and hunt for a viral cause for cervical cancer.

Prophetically, in 1842 Domenico Rigoni-Stern, a surgeon from Verona, Italy, had analysed death certificates and noted that nuns had a lower risk of cervical cancer than married women, and both had a much lower risk than female sex workers.[13] He concluded that development of this cancer is related to sexual contact. So could it be a sexually transmitted virus?

Today, cancer of the uterine cervix causes around 240,000 deaths in women annually, only surpassed by breast cancer. Cervical cancer occurs worldwide but is most common in the countries of Central and South America, sub-Saharan Africa, and Southeast Asia. Finding a cause with a view to preventing these deaths has always been an urgent goal; could it be a virus?

We first met human papilloma viruses as the cause of harmless warts in Chapter 1, and the ingenious way in which they persist in skin stem cells is described in Chapter 6. However, the

HPV family contains many different types, and although most are harmless, a few are decidedly unfriendly.

Back in the 1970s the sexually transmitted herpes simplex type 2 virus was the prime suspect for causing cervical cancer, entirely because almost everyone with the tumour had antibodies to the virus. Yet, using identical molecular techniques to those that identified EBV in nasopharyngeal carcinoma, zur Hausen could not find a trace of HSV-2 in cervical cancer cells.

Some 40 years earlier, Richard Shope at the Rockefeller Institute in New York had heard from a game-hunter that there were lots of rabbits in the State of Iowa with large, warty skin tumours. Having obtained some of these rabbits, Shope managed to transmit the warts to healthy wild rabbits by painting a filtered tumour extract onto their skin. When he applied this extract to the skin of domestic rabbits, some of the warts grew into spreading skin cancers, the cause of which turned out to be a papilloma virus. Zur Hausen may have been influenced by this historical account about rabbits, but he had certainly heard more contemporary anecdotal reports describing benign human genital warts transforming into malignant growths. He decided to follow this up.

Zur Hausen extracted HPV DNA from a verruca (plantar wart) and used this as a probe to search for the same HPV DNA sequences in genital warts. The results were uniformly negative. This was such a surprising result that zur Hausen and his team decided to investigate the possibility of there being more than one type of HPV. And they were right; in fact, the HPV family turned out to be very large, presently containing over 100 different types. With this realization the work took off, and in 1983–4 zur Hausen's research group showed that HPV types 16 and 18 (now known as HPV high-risk strains) were present in 60 per cent and 20 per cent of cervical cancers, respectively.[14]

For this work zur Hausen was awarded the Nobel Prize in Physiology or Medicine in 2008.

HPV was later found to cause other anogenital cancers, throat cancer, and cancer of the penis in men.

Hepatitis B virus (HBV) and hepatitis C virus (HCV) are very different types of virus, but nevertheless, as discussed in Chapter 6, both are spread by blood contamination, cause acute hepatitis, and set up a lifelong persistent infection in around 10 per cent of those infected. This viral persistence causes inflammation of the liver that can lead to chronic hepatitis, cirrhosis, and eventual liver failure. What's more, some 20–50 years after the initial infection liver cancer may develop. These discoveries were made during large epidemiological surveys of global virus prevalence.

The tortuous investigations that led to Baruch Blumberg's discovery of HBV were followed by population screening that uncovered the enormous global burden of persistent HBV. The majority of these infections and deaths occur in Southeast Asia and sub-Saharan Africa, and since liver cancer is among the leading causes of death precisely in these regions, the link between HBV and liver cancer was made. So although HBV was first discovered in 1965, just a year after Epstein discovered EBV, it took a decade to recognize it as a tumour virus.

Similar epidemiological surveys later showed that lifelong HCV infection is also linked to liver cancer. High prevalence of HCV, around 2 per cent, is found in Eastern Europe and the eastern Mediterranean region, where 20–30 per cent of infections are associated with intravenous drug use. But in the Nile Delta in Egypt, where the virus was unwittingly spread by the use of unsterile injecting equipment during a treatment campaign for the parasitic disease schistosomiasis, the overall prevalence reaches over 8 per cent.

The other two known human tumour viruses are human herpesvirus 8 (HHV8) and Merkel cell polyoma virus, both discovered by husband-and-wife team epidemiologist Patrick Moore and molecular biologist Yuan Chang, from Columbia University, New York. The tumours these viruses cause are rare, but the first, Kaposi's sarcoma (KS), reached epidemic proportions during the early stages of the HIV pandemic in the 1980s, when up to one in five HIV-1-positive gay men in the US and Europe developed the tumour (Figure 33).

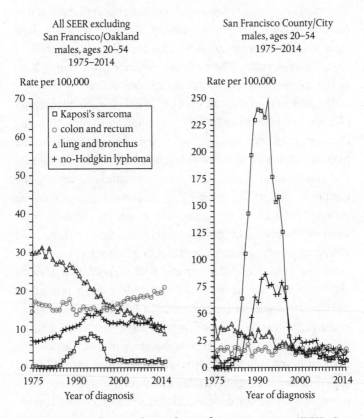

Figure 33. Chart showing the incidence of Kaposi's sarcoma. (SEER: Surveillance, Epidemiology, and End Results).

KS is a tumour of cells that line lymph channels and blood vessels, and it most commonly occurs in the skin, where it affects blood vessels, causing unsightly purple-coloured patches. However, in the later stages of the disease, tumour cells may also invade internal organs.

Interestingly, within the HIV-1-positive population, KS was 20 times more common in homosexual and bisexual men than in those who acquired HIV-1 through infected blood, heterosexual contact, or transmission from mother to child. This suggested that KS was caused by a virus that spread by sexual contact, commonly infected gay men, and was deregulated by the severe immunosuppression of AIDS.

To hunt for the culprit virus, Chang and Moore used a molecular technique called 'representational difference analysis', which detects foreign DNA sequences in tissue samples. In this case 'foreign' means viral, and the DNA they used as a probe was extracted from a KS tumour. They compared this with DNA from the normal skin of the same individual, arguing that the two DNA sequences should be identical unless there was a virus in the tumour cells. In fact, Chang and Moore picked up two DNA sequences which were unique to the tumour material, and these proved to come from an undiscovered, oncogenic herpesvirus. This new virus was named Kaposi sarcoma-associated herpesvirus, and then, more formally, HHV8, being the eighth human herpesvirus to be identified.[15] HHV8 is consistently found in over 95 per cent of KS tumours of all types, and its DNA is present in all the tumour cells.

After this breakthrough Chang and Moore used the same technique to hunt for viruses in many human tumours—mostly without success. Nevertheless, they did find one more human tumour virus—Merkel cell polyoma virus (MCV), which causes the skin cancer called Merkel cell carcinoma.[16] Overall, 86 per cent of these tumours contain MCV, which is a small DNA virus belonging to the polyoma virus family. Several polyoma viruses that infect animals are known tumour viruses, but MCV, the fifth human polyoma virus to be discovered, is the only one that is tumorigenic.

Of the seven human tumour viruses so far discovered, two are herpes viruses—EBV and HHV8—but the other five all belong to different virus families. So what do these viruses have in common that causes them to induce tumours?

There are several characteristics that these viruses share. First, they all set up a persistent infection in humans that may remain for the life of the host. Second, all tumour virus infections are more common than the tumours they cause, some being virtually ubiquitous in humans. Third, tumour development occurs many years after the initial virus infection. And fourth, tumours caused by viruses are clonal, meaning that they arise from a single virus-infected cell.

These observations show that tumour induction by a virus is a rare outcome of a persistent infection. Biologically speaking this makes sense, since it is not advantageous for a virus to cause a tumour that may kill its host, because it would then also kill the virus. Thus, tumour formation is not a part of the virus's normal life cycle, but rather an accidental event, or more likely series of events, that occur in a single virus-infected cell during the long association between the virus and the cell in which it lodges.

Clearly the evolution of a virus-associated tumour is more complex than the simple equation virus infection = cancer. Several other factors are involved in driving just one among many virus-infected cells to tumour growth. To investigate what these factors might be, we must first look at how normal cell growth and division is regulated.

Controlling cell division

Cancer cells are totally out of control. They divide endlessly, regardless of anything else, squashing, invading, and destroying normal tissues as they go. They abide by none of the constraints that regulate normal cells. So what are these constraints? What stimulates a cell to divide, and what causes it to stop again?

The skin is the best example I know of to illustrate cellular growth control because it fits our bodies perfectly—we are born with a compact outer covering and as we grow, it grows with us. It does not get so stretched that we look like an over-inflated balloon or split when we try to move; neither is it so loose that it hangs in folds, resembling the skin of a bloodhound dog. The number of skin cells increases to cover our growing body and if we make a hole in the covering, cells grow in from the sides until the hole is mended and then they stop growing again. How? What controls them?

Cells respond to an array of chemical signals from their surroundings which instruct them when to divide and when to rest. After a skin injury, for example, damaged skin cells release chemicals that bind to surrounding cells and transmit signals to their nuclei. This activates the genes driving cell division, oncogenes, and the cells then divide until the breach is mended. Then tumour-suppressor genes are activated which stop the cells dividing. In normal cells, the actions of oncogenes and tumour suppressor genes are counterbalanced yin–yang style, resulting in finely tuned control.

However, if an oncogene is permanently switched on and unresponsive to normal controls, then the cell will divide continuously. Similarly, if a tumour-suppressor gene is permanently switched off, the effect will be the same—the cell will continue dividing endlessly. A malfunction in any of the chemical pathways leading from the cell surface to these genes in the nucleus may upset the balance, so there are many ways to turn a normal cell into one which is continuously dividing. This is the beginning of the making of a cancer cell, and these are the complex control mechanisms which tumour viruses usurp. But there is still another control mechanism that a cell can impose to prevent outright anarchy.

A group of genes that are vital for survival regulate a self-destruct mechanism called 'programmed cell death' or 'apoptosis'. This process is essential for ridding the body of harmful viruses, since death of a virus-infected cell before it produces thousands of new viruses can prevent an infection spreading throughout the body and killing the whole organism. Apoptosis may also be triggered if cell division gets out of control or if the cell's DNA is so badly damaged that it is a danger to the body as a whole. This so-called suicide programme activates cellular enzymes which chop up DNA, causing instant cell death. In just 30 minutes the cell is irrevocably damaged and is rapidly gobbled up by a passing macrophage.

So how do viruses influence these cellular genes to cause a tumour? The progression from virus infection to tumour development involves a slow accumulation of abnormalities, each of which is in itself rare. This step-by-step evolution explains not only why just a minority of people carrying a particular virus develop the associated cancer, but also why the tumour results from a single cell, and why there is often a very long interval between initial infection and cancer development. With liver cancer, for example, up to 50 years may elapse between HBV infection and the appearance of cancer, so among HBV carriers liver cancer is more common in those who were infected early in life, particularly those infected during, or shortly after, birth. The high number of HBV carriers in Southeast Asia means that transmission of virus from mother to child around the time of birth is common, and therefore the incidence of liver cancer in this area is very high.

Retroviruses like HTLV-1 integrate into the cellular DNA. From this position, the HTLV-1 gene called *Tax* is expressed, and this plays a central role in oncogenesis. Similarly, the DNA of both HPV and HBV integrates into the cellular DNA chain, but in these cases this is not part of the virus's normal life cycle. The integration represents a mistake that occurs in a single virus-carrying cell during cellular DNA replication, and once it occurs the virus's oncogenes are permanently switched on.

The large DNA genomes of the two oncogenic herpesviruses, EBV and HHV8, do not integrate into cellular DNA but sit in the cell nucleus as a circular plasmid. Both viruses carry genes that can inhibit activation of the immune response and also act as viral oncogenes, interacting with cellular genes that induce cell proliferation and inhibit apoptosis. EBV targets blood B cells, and when normal B cells are infected with the virus in the laboratory, it causes the cells to grow into cell lines that proliferate indefinitely; a powerful indication of the virus's ability to take control of the cell cycle.

As the first virus-associated human tumour to be discovered, now more than 50 years ago, much has been elucidated about the steps involved in the evolution of Burkitt lymphoma (BL) from a normal B cell. In BL only one viral gene is expressed, and the resulting protein is essential for maintaining the viral DNA in the cell and also enhances cell survival by inhibiting apoptosis. But clearly this is not enough to cause BL, so there must be additional events that push an infected cell towards malignancy.

We know that EBV is very common, infecting around 95 per cent of the world's adult population, and virtually 100 per cent in those over two years of age in the regions where BL occurs. Back in the 1960s, Burkitt recognized holoendemic malaria as a cofactor in tumour development, identifying this link as the reason for the tumour's clear geographical restriction to equatorial Africa and Papua New Guinea. In these areas almost all children are constantly infected with malaria parasites as solid immunity to malaria is not established until around the age of 12 years.

Exactly what contribution malaria infection makes to the evolution of BL is not clear, but recurrent malaria infections certainly cause immune suppression on one hand and proliferation of B cells, sometimes on a massive scale, on the other. These effects together probably account for the fact that during an acute attack of malaria, children living in tropical regions have high levels of EBV-infected B cells in their blood, so increasing the chances of further malignant change in one of these cells.

But since both EBV and malaria infections are virtually universal in BL endemic areas, scientists knew that there must be at least one more event required to induce the tumour, this time affecting just a single cell, finally releasing it from all growth constraints. And in 1972 scientists at the Karolinska Institute in Stockholm, discovered it: a chromosome switch, or translocation. This is a rare genetic accident that occurs during cell division in which a piece of chromosome 8 is accidentally relocated to chromosome 14. As a result, the cellular oncogene located on chromosome 8, called *c-myc*, is permanently switched on, causing uncontrolled cell division and the development of BL.[17]

Immunosuppression

It is no surprise to find that tumours caused by viruses are more common in people whose immune system is suppressed than in the general population. But while this used to be a rare event, the HIV-1/AIDS epidemic in the 1980s reinforced the link. Also, numbers of immunocompromised people have increased enormously because of the success of immunosuppressive treatments used for many conditions, most commonly after organ transplantation and for cancer treatment. As we noted previously, transplant recipients have particular problems with persistent viruses which are either already lodging in their body before the transplant or may actually be delivered inside the donor organ. While the success of immunosuppressive drugs in preventing graft rejection has increased the scope of organ transplantation to include not only kidneys, liver, heart, and lungs but also trachea, bowel, and pancreas, there is always a difficult balance to be maintained between the dose of immunosuppressive drugs that protect the transplant from rejection and the risk of a malignancy caused by a tumour virus.

Particularly common in this regard is a lymphoma called B lymphoproliferative disease caused by EBV. This represents an outgrowth of EBV-infected B cells in the absence of adequate immune control of the persistent infection. In this tumour, unlike BL, the cells do not carry the 8:14 chromosomal translocation. But the malignant B cells express all the EBV latent genes, including the viral oncogenes, and it is these that drive cell proliferation. And because the latent genes are expressed in every cell, this tumour responds well to immunotherapy.

Tumour therapy

Finding that a virus is involved in the outgrowth of a tumour opens up the possibility of preventing tumour development by preventing the infection in the first place. The easiest way to do this is with a vaccine, and this method has been particularly successful with HBV and HPV.

At first, the prospect of preventing HBV infection by means of a vaccine looked bleak. Although a safe and effective vaccine had been available since 1982, initially it was too expensive for governments and international organizations such as the WHO to provide it to countries in sub-Saharan Africa and the Far East, where it was most needed. But thanks to an international task force of influential scientists pressurizing vaccine manufacturers to reduce the price, and convincing governments of the seriousness of the problem, HBV vaccination programmes are now in place in more than 110 countries. The aim is to prevent infection by immunizing all newborn infants, and countries where the programme is up and running have seen a dramatic drop in persistent HBV infection. Nevertheless, there is more to do. The present vaccine requires a course of three injections over six months, and even then there are non-responder levels of around 10 per cent. Presently these problems disproportionally impact on countries where levels of HBV infection and associated liver cancer are highest.

A vaccine against HPV types 16 and 18 became available in 2006, and this gives almost 100 per cent protection against persistent infection with these two high-risk virus types. Like the HBV vaccine, it is expensive and difficult to administer in developing countries where it is most needed. However, presently 115 countries have introduced an HPV vaccine programme, generally giving it to girls at the age of 14 years. Additionally, a few countries, including the UK, have adopted a gender-neutral programme, immunizing boys alongside girls to prevent sexual spread of the virus, and also as protection against anogenital, throat, and penile cancers. Figures from the UK show that vaccination has reduced HPV 16 and 18 infection in women by

90 per cent, with a similar reduction seen in high-grade, precancerous lesions of the cervix.

Where virus-associated tumours occur in people with immunosuppression it is sometimes possible to restore the failing immune system. In organ transplant recipients, for example, this may be achieved by reducing the dose of immunosuppressive drugs and allowing killer T cells, which recognize and kill virus-infected cells, to act normally again. However, doctors often find themselves in a catch-22 situation—give too much immunosuppression and allow the tumour to grow; or give too little so that the tumour may shrivel and die, but the vital transplanted kidney, heart, lung, or liver is jeopardized.

Using a technique pioneered by researchers at St Jude's Hospital in Memphis, Tennessee, clones of killer T cells that only recognize and kill EBV-infected cells can be grown in the laboratory,[18] but unfortunately, in order to function, the T cells, like the transplant itself, have to be matched to the recipient's tissue type. This presently makes the process too time-consuming and expensive for routine treatment. However, killer T cells remain viable when frozen, and researchers are developing frozen banks of these T cells that can be thawed on demand.[19] So the time may come when 'spare part' surgery is commonplace and each recipient has killer T cells recognizing all the dangerous viruses and matched to their own tissue type stored before the transplant surgery, to be used to counteract infections at a later date.

For those tumour viruses for which no vaccine is yet available, antiviral drug treatments are being used with some success. These and other ways of combatting viruses are discussed in the next chapter.

8

TURNING THE TABLES

Smallpox was the most lethal of the recurrent childhood infections and, until the late eighteenth century, had it all its own way. But in 1715 when smallpox virus infected the beautiful young Lady Mary Wortley Montagu (1689–1762, Figure 34), the fightback began. Although she survived the smallpox, it left her face pock-marked and her eyes devoid of lashes. This turn of events gave Lady Mary a keen interest in smallpox that led, a few years later, to the first successful prevention of the disease in Europe.

Lady Mary Wortley Montagu (1689–1762)

Lady Mary's husband, Sir Edward Wortley Montagu, was appointed British Ambassador to Turkey in 1716 and they travelled together to Constantinople, where they lived for the following two years. Here Lady Mary saw smallpox prevention in practice and wrote enthusiastically to her friend, Sarah Chiswell:

> The smallpox, so fatal, and so general among us, is here entirely harmless by the invention of *engrafting*, which is the term they give it. A set of old women perform the operation by scratching open a vein in the patient and putting into it as much of the

smallpox venom as could lie on the head of a needle. On the eighth day the patient ran a fever which kept him in bed two or three days, and afterwards seldom showed pockmarks.[1]

Lady Mary was referring to the practice of 'variolation' or 'inoculation', which is thought to have been adapted from the practice first recorded in China in the 1500s where it was accomplished by inhalation of material from cases of smallpox. In Turkey in the 1700s, healthy people were inoculated with scrapings from smallpox pocks through a scratch in the skin, after which they usually suffered a mild infection followed by lifelong

Figure 34. Picture of Lady Mary Wortley Montagu (1689–1762).

immunity. Lady Mary was so convinced that this was a safe and effective way to prevent smallpox that in 1717 she had her six-year-old son inoculated. Seven days later he developed a fever and about a hundred pocks. But the spots crusted and fell off leaving no scars; the experiment had been a complete success.

Then, in 1721, shortly after Lady Mary's return to England, a severe smallpox epidemic raged in London, and she used her position in high society to encourage the use of inoculation. But, as the following quote from one of her letters indicates, she clearly had little faith in the medical profession helping her in this enterprise because of the income they stood to gain from a smallpox epidemic:

> I am patriotic enough to take pains to bring this useful invention into fashion in England, and I should not fail to write to some of our doctors very particularly about it, if I knew any one of them that I thought had virtue enough to destroy such a considerable branch of their revenue, for the good of mankind. But that distemper is too beneficial to them not to expose to all their resentment the hardy wight that should undertake to put an end to it.[2]

That same year Lady Mary's daughter was successfully inoculated in the presence of several influential doctors and so the practice became more widely known. Caroline, Princess of Wales, was interested but cautious. Before committing her two daughters to this new safeguard against smallpox she had it tested on six condemned prisoners in Newgate prison. They volunteered for inoculation on the understanding that if all went well they would be released, and so it turned out—although the inoculation caused no symptoms at all in two of the prisoners, probably because they were already immune.

But the Princess was still not satisfied, so she arranged for 12 orphans from the Parish of St James's, Piccadilly, to be inoculated at her own expense, before eventually allowing her daughters, the Princesses Amelia and Caroline, to be treated in 1723. The procedure was successful and thereafter the practice of inoculation spread in the UK and the US. Between 1721 and 1728, 897 people were inoculated in Britain and 17 died of the effects. But during the same period more than 18,000 died of smallpox, so, if all those surviving inoculation were immune to smallpox, it seemed the safer option.

However, inoculation was obviously not entirely safe and was not universally accepted. Some doctors saw it as the wanton spreading of infection and, since at the time disease was widely thought to be a punishment from God, many of the clergy believed that it interfered with God's will. Despite this, it continued to be popular until 1798 when Edward Jenner published the details of a safer alternative.

Edward Jenner

Jenner was an English country surgeon–apothecary who lived and worked in Berkeley, Gloucestershire. In this rural setting Jenner became interested in cowpox, an infection of cows' udders that sometimes spread to the hands of dairy workers, causing localized blisters and a mild fever.

In his routine daily medical practice Jenner performed many inoculations against smallpox, and he noticed that when inoculated, dairy workers often produced no response. In fact, whether or not they had suffered from smallpox in the past, they reacted as if they were already immune. Jenner deduced that cowpox infection must protect dairy workers from smallpox and he decided to test out this theory in a remarkably direct and practical way. In 1796 Jenner inoculated James Phipps, the eight-year-old son of his gardener, with live cowpox obtained from the hand of a local milkmaid called Sarah Nelmes (Figure 35). Then, some weeks later, he tested the boy's immunity by inoculating him with live smallpox (Figure 36). Fortunately, the child remained healthy, indicating that he was indeed immune to smallpox. In Jenner's own words:

> During the investigation of the casual Cow Pox, I was struck with the idea that it might be practicable to propagate the disease by inoculation, after the manner of the Small Pox, and finally from one human being to another. I anxiously waited some time for an opportunity of putting this theory to the test. At length the period arrived. The first experiment was made

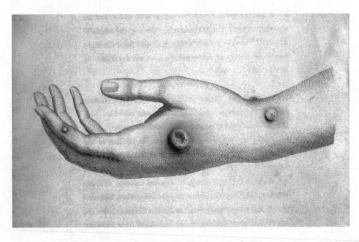

Figure 35. Hand of Sarah Nelmes showing cowpox pocks.

Figure 36. Bronze statue by Guilio Monteverde showing Edward Jenner vaccinating a small boy.

upon a young lad of the name of Phipps, in whose arm a little Vaccine Virus was inserted, taken from the hand of a young woman who had been accidentally infected by a cow. Notwithstanding the resemblance which the pustule, thus excited on the boy's arm, bore to variolous inoculation, yet as the indisposition attending it was barely perceptible, I could scarcely persuade myself the patient was secure from the Small Pox. However, on his being inoculated some months afterwards, it proved that he was secure.

This historic human experiment, which was based on astute practical observation combined with logical deduction, cut across all ethical and theoretical barriers to produce one of the most remarkable and groundbreaking medical achievements ever recorded.

At first, certain members of the medical profession argued against vaccination, maybe for the pecuniary reasons alluded to earlier by Lady Mary in her letter to Sarah Chiswell, but, thanks perhaps to Lady Mary's pioneering work which had set the scene, its obvious benefits were soon appreciated. Mass vaccination programmes were organized in Britain and Jenner's fame spread. In 1805 Napoleon gave the order for all his forces to be vaccinated and, in 1806, US President Thomas Jefferson wrote to congratulate Jenner:

> Yours is the comfortable reflection that mankind can never forget that you have lived. Future nations will know by history only that the loathsome small-pox has existed and by you has been extirpated.[3]

Neither Lady Mary nor Jenner knew why or how their particular method of smallpox prevention worked, but both methods relied on producing enough infection to stimulate an immune response without causing severe disease. Lady Mary's inoculation succeeded because a small dose of smallpox virus was administered by an unnatural route. Smallpox virus inoculated through the skin, rather than by the natural airborne route straight to the lungs, gets off to a slow start. By the time it spreads to a systemic infection, antibodies and immune T cells are ready to control and eliminate it. The procedure depended on material known as 'scabs' collected from the pocks of active smallpox cases and used directly for the inoculation. Even in the most careful hands this was a risky procedure because scabs were certainly capable of causing full-blown smallpox. Indeed, as we saw in Chapter 1, in 1763, during the fight for new territories in America, blankets deliberately contaminated with 'scabs' were distributed to Indigenous Americans with devastating effect.[4]

The success of Jenner's cowpox vaccination in protecting against smallpox relies on two facts. First, cowpox and smallpox viruses are very similar—their genetic material is 95 per cent identical, and this is sufficiently close to trick the human immune system into recognizing them as one and the same thing. Second, cowpox virus only causes a mild localized disease in humans, and in doing so induces lifelong immunity to itself and gratuitously to the smallpox virus as well. Smallpox vaccination was the first example of an attenuated or slightly different strain of virus, in this case cowpox, used to protect against its more dangerous relative, predating the later commonly used strategy by over a hundred years.

The original cowpox virus used by Jenner came direct from the hand of a milkmaid with the infection, but following this, direct arm-to-arm inoculation was used to pass the virus from an inoculated person to a non-immune individual. Later still, another strain of pox virus—vaccinia—was introduced because it could be grown on the skin of calves, harvested, and used for several vaccinations. Vaccinia virus is the parent of the virus still used today if smallpox vaccination is required, but its origins and the identity of its natural host have been lost in the mists of time.

Although vaccination soon became widespread in the western world, where it had a dramatic effect in preventing smallpox, this was not the case everywhere. In poorer countries, difficulties in organizing vaccine programmes, combined with the instability of the vaccine at high temperatures, meant that smallpox remained the world's number-one killer virus, causing at least 500 million deaths in the 100 years before the successful outcome of the WHO smallpox eradication programme in 1980.

Louis Pasteur

Following smallpox, rabies virus was the next to be prevented by a vaccine, this time produced by microbiologist Louis Pasteur working in Paris in the mid 1800s. Although Pasteur was among the first to identify bacteria in diseased material, like Jenner, he had little idea of the microbe he was dealing with when he set out to make a vaccine against the deadly rabies virus. To prepare the vaccine he used spinal cords from rabies-infected rabbits and inactivated the virus they contained by leaving the material to dry out in a desiccator. After 14 days all the infectivity was destroyed, and he then injected dogs on 14 consecutive days with the rabbit spinal cord preparations dried for increasingly shorter periods. Two weeks after the last injection he challenged the dogs with fully active virus and found that they were immune to rabies.

In 1885 Pasteur reckoned he was about two years away from having a rabies vaccine safe for human use, when he was persuaded to try his preparation on a nine-year-old boy from Alsace called Joseph Meister. The boy had been severely bitten by a rabid dog two days previously and there was no other hope for him. So Pasteur agreed to give him the same course of injections as he had given to the dogs, with the last dose being the fully virulent rabies. Like the dogs, the boy survived the injections and did not develop rabies. This success made Pasteur's vaccine famous and people bitten by rabid animals came to him for treatment from all over Europe.

Vaccines for all

From the mid 1950s onwards a surge in vaccine production saw common viruses like polio, measles, rubella, and mumps, as well as common bacterial infections like diphtheria, pertussis, and tetanus, being rolled out to all children in western countries. These vaccines have been so successful in preventing the acute childhood virus diseases that those born in developed countries after the mid 1970s are unlikely to have suffered from any of the them (with the exception of chickenpox, which is caused by a persistent virus), and the fear of paralytic polio has become a distant memory (Figure 37). Here we look at how these vaccines were prepared and the recent advances in vaccinology spurred on by the COVID-19 pandemic.

Traditional vaccines were made in two ways. 'Live' vaccines used viruses that had been weakened or attenuated, usually by prolonged growth under unfavourable conditions, while killed vaccines used chemically treated, 'killed' viruses that were completely non-infectious.

The success of these vaccines leaves no doubt that immunization has had an overwhelmingly beneficial effect on global human health. It has reduced disease burden in every country in the world and was the key to eliminating wild smallpox from the entire planet. But vaccines will never be completely safe, so these immense achievements have brought their own problems.

Vaccines produced before the early 1900s, like rabies vaccine, sometimes caused severe allergic reactions because they were made in animals and were difficult to purify and standardize. Then, later, viruses for vaccines were grown in hens' eggs. This improved things slightly, but nevertheless they were not entirely safe because they still contained animal proteins. These problems

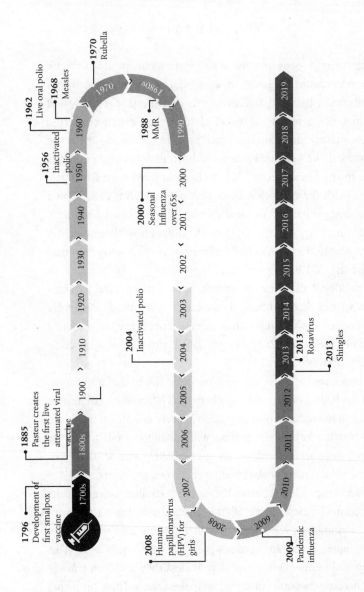

Figure 37. Historical development of major vaccines and the introduction of routine vaccine programmes in the UK. MMR = Measles, mumps and rubella triple vaccine.

were mainly overcome when cell culture was introduced for large-scale virus production—but this brought other concerns.

Between 1955 and 1963, polio virus for vaccine production was grown in cultured rhesus monkey kidney cells. But in the 1960s, a new simian virus, called SV40, was isolated from monkey kidney cells and was quickly found to contaminate polio vaccine. Although chemical treatment effectively inactivated the polio virus in the vaccine, it did not always kill SV40. Tests showed that live SV40 contaminated up to 30 per cent of vaccine batches and people immunized with these batches tested positive for antibodies to SV40, indicating that they had been infected with the virus.

The problem was compounded when SV40 was found to cause tumours in animals, so many people around the world had inadvertently been infected with a potential tumour virus—98 million in the US alone. Groups of immunized people were monitored for increased tumour development for years afterwards, but thankfully none was found. And the realization that cultured animal cells can harbour potentially dangerous viruses led to the use of human cells for vaccine production wherever possible.

Live attenuated vaccines usually induce more long-lasting immunity than killed preparations since they actually grow in the body, albeit briefly. But because of this, they have their own particular shortcomings. They cannot be used during pregnancy for fear of infecting the developing baby, and if given to patients with immune deficiency they may spread and establish a generalized or persistent infection. Also, mutations causing reversion from attenuated back to a virulent form of virus can occasionally occur.

The issue of reversion was highlighted by the live attenuated polio vaccine that replaced the killed polio vaccine in the 1970s. The live vaccine had the great advantage of being given orally rather than by injection, and because it infects cells lining the gut and is excreted in faeces in large amounts, it often had the knock-on effect of spreading the vaccine virus to family members via the faecal–oral route, so boosting herd immunity.

Although this live attenuated polio vaccine is extremely safe, it can revert to the wild form, and as such causes paralytic polio in around one in every 2 million of those vaccinated. This level was deemed acceptable when wild polio was an enormous threat, but as the wild virus became rare, so vaccine-induced disease became relatively more common. In addition, elimination of polio from a country or continent could not be achieved using the live vaccine that circulated freely in communities and occasionally reverted to the wild form. These observations necessitated a change back to the killed vaccine preparation for the final stages of the WHO polio eradication programme, which has been successful in eliminating polio from most of the world. Presently only Pakistan and Afghanistan have endemic wild polio, while several countries still have the vaccine strain circulating.

Vaccines—victims of their own success?

With the diminishing likelihood of catching a particular virus, the small chance of suffering one of the rare complications of the vaccine that prevents it becomes less acceptable. Thus, for a single individual, logically there comes a time when it may be safer *not* to be vaccinated. This was certainly the case for small-pox, well before it was officially declared eradicated. Yet, if vaccination rates for any other virus fall, then a significant non-immune population will emerge and epidemics will return. So, there is a conflict of interests between protection at a personal and at a population level. Each individual is dependent on others in the community to maintain a high corporate level of immunity, that is, herd immunity. At the level of national public health services, mass vaccination is relied upon to keep their populations healthy and consequently vaccine safety is paramount. Over the years there have been many vaccine scares—some real, others just rumours—but all potentially dangerous if they discourage people from being vaccinated.

In 1998, a scare over the use of the measles/mumps/rubella triple vaccine (MMR) in the UK reverberated around the world, and has since created a problem with serious global health implications. The issue began with a report published in the medical journal *The Lancet* suggesting an association between MMR—particularly the live measles vaccine component—chronic bowel disorders, and the behavioural disorder autism.[5] The report received massive media coverage at the time and caused an immediate fall in MMR uptake. Despite scientific experts quickly refuting and later disproving the claimed associations, 2,000 fewer children than usual were immunized in the following three-month period in the UK. The senior author of the report was found guilty of dishonesty and of flouting ethics protocols, and was struck off the UK Medical Register. But the problem did not end there. He simply moved his operation to the US, where he found fertile ground for his anti-vaccination propaganda. This has since morphed into an international anti-vax movement that has taken to social media to spread what the vast majority of experts regard as misinformation.

An estimated 95 per cent vaccine coverage rate is required to completely arrest the spread of the measles virus (Ro 12–18). The result of falling uptake of measles vaccine was the inevitable resurgence of measles in the UK, the US, and several European countries from 2017 onwards, and in 2019, the UK lost its WHO 'measles-free' status. All this has caused unnecessary deaths and has hampered the WHO measles eradication programme.

Protection against flu

Flu is now one of the commonest infectious diseases worldwide, causing outbreaks virtually every winter and pandemics at around 10–40-year intervals, and yet it has proved very difficult to control. As soon as flu virus was successfully grown in the laboratory in the early 1940s it was used in an inactivated form to produce a vaccine, and vaccination has remained the mainstay of protection against flu ever since. But there are problems, primarily due to the unpredictability of flu virus's propensity to genetic drift and shift.

With this in mind, as long ago as 1947, the WHO set up the Global Initiative for Worldwide Flu Virus Surveillance, and this still exists, albeit in an extended form today. It comprises 143 institutes in 113 member states, with each national institute monitoring flu strains circulating in their particular region, identifying and reporting new strains back to one of the six WHO Collaborating Centres situated in Australia, China, Japan, the UK, and the US (2) (Figure 38). Decisions on the composition of the annual flu vaccine are based on information generated by this global network.

Up to three strains can be included in the annual vaccine without affecting the vigour of the immune response to each, but to be effective the cocktail has of course to include the coming winter's strain, and this requires a certain amount of guesswork. However, since almost all new flu viruses have their origins in the Far East, where they circulate among animals, mainly ducks, kept in cramped conditions on farms and in markets, the vaccine's composition is based on those strains

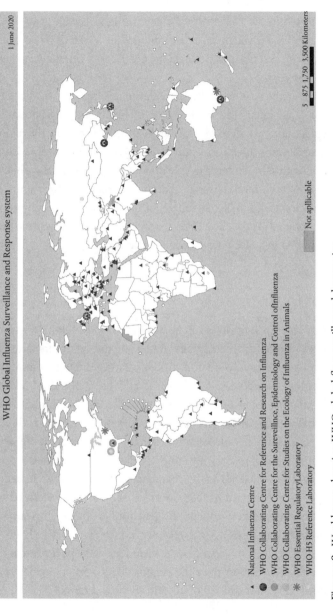

Figure 38. World map showing WHO global flu surveillance laboratories.

detected in the Eastern hemisphere in the preceding season. In the UK, the trivalent vaccine is offered free to the over-65s every autumn, and this practice cuts the death rate from flu by around 50 per cent.

Since the year 2000, flu surveillance in China and the Far East has been intensified in the hope of identifying new potential epidemic or pandemic strains of flu as they arise. Now the region boasts more than 400 laboratories dedicated to flu virus isolation with samples feeding in from 554 sentinel hospitals, while at the research level, farmers, abattoir workers, and their animals are monitored for likely emerging new strains. As discussed in Chapter 4, just such an event occurred in 1997 when bird flu emerged in a Hong Kong poultry market.

Subunit vaccines

Not all virus infections can be prevented by traditional vaccines made from whole microbes. But from the 1980s onwards, as viral genes that code for proteins targeted by the immune response were identified, these could be synthesized in the laboratory for use as 'subunit' vaccines. The key proteins are usually virus surface receptor molecules, since antibodies directed against them, called neutralizing antibodies, prevent the virus from binding to and infecting a cell. The first of these new-generation vaccines to be manufactured was against hepatitis B virus, introduced in 1982. In this case, yeast cells were genetically engineered to contain the gene for hepatitis B virus surface protein, part of the virus receptor complex. Yeast cells grow easily in culture to produce large quantities of the pure protein—a much safer option than the original HBV vaccine, which was purified from the blood of hepatitis B virus carriers.

Other new-generation vaccines use the genes that code for target proteins cloned into harmless carrier viruses such as poxviruses, adenoviruses, and herpesviruses. When inoculated, the carrier virus infects cells and induces them to express the vaccine protein on the cell surface, so stimulating the desired immune response. Other techniques discard the harmless carrier virus and use naked DNA or RNA sequences from essential virus genes (see below for COVID-19 vaccines). But despite all these technological advances, on occasions it seems impossible to produce a vaccine against certain viruses. Presently included on this list are the common cold virus, SARS, MERS, and, most importantly, HIV-1.

Why no HIV-1 vaccine?

There is little doubt that a vaccine against HIV-1 would be the best and most cost-effective way to curb the pandemic, which has been ongoing for more than 40 years. But despite massive financial input, scientific effort, and many clinical trials, none of the prototypes has been effective.

At first, scientists were optimistic about producing a vaccine fairly quickly. After all, HIV-1 only has a handful of genes, so it should be easy to work out which are targets for the neutralizing antibody response. However, HIV-1 is a devious customer and many unforeseen problems arose. One of these is the extreme variability of the virus; its unique flip from RNA to DNA after infection of a cell is a highly error-prone procedure which generates many mutants, some of which outflank the immune response. So not only are viruses in one infected individual always changing, but viruses from different people are very diverse. This makes it difficult, if not impossible, to design a single vaccine that will protect everyone from infection.

Most potential vaccines target the HIV-1 gp120 protein, the receptor molecule that latches on to the cellular CD4 molecule to initiate infection. Many prototype vaccines successfully generated antibodies against gp120 but still failed to protect against infection. And if these did not succeed then it is hard to see what will prevent the virus infection. It may be that the persistent infection established by HIV-1 is the problem. Most vaccines prevent disease but do not necessarily produce sterile immunity; that is, they don't completely prevent infection. So even in those immunized against a particular virus, this virus may enter the body silently and infect cells before it is rapidly eliminated having caused no disease. But in the case of HIV-1 infection, insertion of the viral genome into the cell's DNA is an early step in the infection process, and thereafter the virus is hidden from the immune response and cannot be removed. For these reasons, the mainstay of HIV-1 management turned from prevention to treatment with antiviral drugs, and, as we will see later, this has been a phenomenal success in preventing disease progression.

Clinical trials

Once a potential vaccine has been manufactured it must be tested in a standard series of clinical trials before it can be licensed for general use. Prior to the emergence of COVID-19 in 2019, the shortest time ever recorded for completion of this process was four years, for mumps vaccine in 1970, but generally it took around ten years. The potential vaccine must be tested first in laboratory animals, and then in a series of phased human trials labelled 1, 2, and 3.

Phase 1 trials are all about safety, and so just a small number of healthy volunteers are enrolled, vaccinated, and checked for any unacceptable side effects. Success here leads to a larger phase 2 trial, generally with around 100–200 participants, but still no control group. This again identifies common side effects of vaccination like pain and swelling at the injection site and/or generalized, short-lived, flu-like symptoms. Phase 2 trials also determine the doses of vaccine and the number of shots to be tested out in the phase 3 trial. In addition, during phase 2 trials, scientists monitor the immune response to the vaccine to ensure that it induces both neutralizing antibody and T-cell responses that would be expected to protect against the virus infection.

If these preliminary trials give satisfactory answers then a large phase 3 field trial can begin, usually aiming to enrol tens of thousands of participants. Phase 3 trials are randomized, double-blind, controlled trials, and are carried out in an area where the virus under study is spreading. Volunteers should include male and female healthy adults of all ages and ethnicities, and these are randomly allocated to receive either a shot of vaccine or a placebo, with neither the organizers nor the participants knowing which type of shot anyone receives. The organizers will have calculated the number of cases of the disease required to give them a statistically significant result indicating whether the vaccine is effective or not, and from then on it is a waiting game. How long the wait is obviously depends on how common the disease is in the trial area, and this is why large trials often run consecutively on several sites, and also why it may take many years to test a newly developed vaccine against a rare or emerging infection.

Once the predetermined number of cases is reached then the trial code can be broken and the percentage protection, or vaccine efficacy, calculated. The final stage is licensing the product for use in humans, another step that could take years. But all this changed after COVID-19 appeared.

Speeding up but not cutting corners

As soon as COVID-19 emerged in December 2019, vaccinologists around the world were galvanized into action. Within a few months well over 100 vaccines were in production, using all the different techniques described above as well as some new ones. Around 40 vaccines reached clinical trials inside a year, and by the beginning of 2021 several were being rolled out in a desperate attempt to save lives. This is an amazing achievement, and a wonderful legacy for future vaccine production against recalcitrant microbes. So how was it possible?

First, new technologies using viral messenger RNA that carries the genetic code for a potential vaccine had been researched for the previous 10–15 years in attempts to create vaccines for emerging infections like Ebola, Zika, SARS, and MERS, which have proved refractory to traditional methods. In 2019 all this groundwork came to fruition and was ready to go with the new pandemic virus. These vaccines can be synthesized in days in the laboratory, and scale-up poses no problem.

Second, clinical trials are often hampered by lack of funds, but in this frenetic scenario unprecedented amounts of cash were thrown at the projects so that they could steam ahead.

Third, with cases of COVID-19 escalating globally, there was no shortage of volunteers entering clinical trials, and as multiple trial sites were established the waiting time to accumulate statistically significant results was remarkably short.

Fourth, in the rush to get vaccines into the clinic, phase 1, 2, and 3 trials were telescoped into a single enormous trial with the phases effectively running in parallel rather than in series (Figure 38).

Fifth, mass production of vaccines began before the final trial results were in, so that when the go-ahead was given the jabs could be prepared immediately for the mass vaccination programmes.

When the results of the first four trials to reach fruition were revealed, remarkably, they gave protection against severe COVID-19 to levels of between 65 and 95 per cent, and when the regulatory authorities completed their deliberations in a few weeks rather than months or even years, vaccine rollout immediately got under way.

An evolving target

Preventing the initial interaction between a virus and its target cells is key to a successful vaccine. Consequently, most candidate vaccines against COVID-19 were designed to target the SARS-CoV-2 spike protein, the virus receptor that binds to a cell. Among the first to hit the clinic in the UK were two vaccines that both used spike protein RNA, one with the mRNA enclosed in a lipid envelope, the other utilizing a harmless chimpanzee adenovirus to carry spike protein RNA into cells (Figure 39). All good so far, but at the time of writing there are still some unknowns about these first-generation vaccines that will affect the long- and medium-term success of the COVID-19 prevention programme.

Presently it seems that the new vaccines do not entirely prevent infection and onward transmission, and although so far they are effective against newly arising variants, we do not know how long this and the immunity they confer will last. Nevertheless, the indications are that the new vaccines are proving effective across all age groups, and the impact on death rates, especially in the elderly, has been dramatic where vaccine roll-out has been conducted efficiently.

As an RNA virus, it was always assumed that SARS-CoV-2 would mutate, although not as rapidly as HIV-1, and this has

The virus RNA that codes for the spike protein is coated in lipid so that it can enter a cell – this is the vaccine

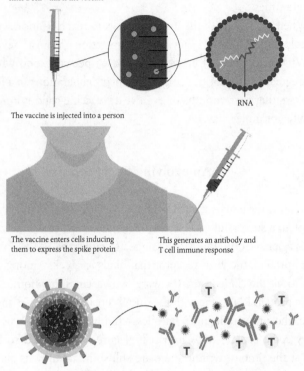

The vaccine is injected into a person

RNA

The vaccine enters cells inducing them to express the spike protein

This generates an antibody and T cell immune response

If the person later encounters SARS-CoV-2 the antibodies and T cells will fight the virus

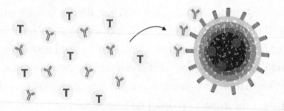

Figure 39. How an RNA vaccine works.

proved to be the case. Although virus RNA is mutating all the time, at the beginning of an outbreak when no one is immune there is no pressure on the virus to alter its makeup and so no particular mutant gets the upper hand. But as immunity increases in a community, mutants with a selective advantage, such as spreading more easily between hosts, start to emerge. And that is just what happened.

In the UK, SARS-CoV-2 RNA sequences had been monitored since the beginning of the pandemic by COG-UK, the COVID Genomics UK Consortium. It is mutations in the spike gene that are most significant since this controls the virus's ability to bind to and infect cells. In December 2020, COG-UK reported 'a variant of interest' with multiple mutations in the spike protein, now referred to as the Alpha variant (B1.1.7). These changes allow the virus to spread more easily between people, becoming 50 per cent more infectious and raising R0 to 4–5. Following its emergence, this variant spread rapidly in the UK and then to more than 50 countries, replacing its less well-adapted siblings. Around the same time as the emergence of the Alpha variant, several others came to light, including those originating in South Africa (Beta), Brazil (Gamma) and India (Delta), all of which spread internationally.

There is obvious concern regarding variants that can evade the protective immunity produced by vaccination, and indications that some vaccines may be less effective against the Beta variant led to many countries enforcing quarantine regulations at borders and severely restricting international travel. But such restrictions cannot be a long-term solution. At some stage we have to accept that COVID-19 is here to stay, for the medium term at least. Achieving herd immunity will be difficult with the current batch of vaccines, and variants will crop up that erode their effectiveness. This all sounds very depressing, but there are ways forward.

In some ways the COVID-19 pandemic is comparable to the 1918 'Spanish' flu pandemic that eventually killed up to 50 million people globally. And like flu, COVID-19 will not go away. It is likely to become endemic, and circulate continuously in the human population. Like seasonal flu, which kills around 250,000 globally every year, COVID-19 will likely cause recurrent epidemics among the non-immune for years to come. *If* infection establishes long-term immunity, then the virus will infect young children and in doing so will become less lethal as its present targets, the elderly and chronically sick, will already be immune. If not then we are in for the long haul, needing to develop and maintain a surveillance network and annual vaccine production scenario akin to that of flu.

The first generation vaccines that prevent severe disease have been a godsend to health services around the world that struggled to cope with the onslaught of severe COVID-19 cases, and have allowed other urgent medical cases with non-COVID-19 conditions to be treated. And the new technology that produced vaccines with such impressive speed can be used to tweak the present vaccines into protecting against virus variants. Realistically, this must be the way forward, and, following in the path of flu prevention, it is eminently possible that a state of normality will be reached, even if not a world without COVID-19.

Microbial treatments

The famous story of Scottish bacteriologist Alexander Fleming's discovery of penicillin began in 1928 when he returned to his laboratory at St Mary's Hospital, London, after a holiday. On examining a pile of neglected bacteriological plates on which he had been growing the bacterium *Staphylococcus*, he noticed a plate contaminated with mould. What drew his attention was that the area surrounding the mould was devoid of *Staphylococcus*, while these bacteria grew happily on all the non-contaminated plates. This observation sparked the discovery of the first antibiotic, penicillin, produced by the mould. This was later purified by Howard Florey and his colleagues in Oxford ten years later, and earned Fleming and Florey the Nobel Prize in Physiology or Medicine in 1945.

The antibiotic era had arrived, and has since saved billions of lives. But although antibiotics are brilliant at curing infections caused by bacteria, unfortunately they have no effect on virus infections. Indeed, because viral reproduction is so intimately involved with the cell it infects, many scientists thought that any drug that disrupts one would necessarily disrupt the other, and so selectively killing viruses was an impossible goal. However, more recently this pessimism has been proved wrong, and now many safe antiviral drugs are available. Nevertheless, these are not the panacea which broad-spectrum antibiotics have proved to be against bacteria, because on the whole antiviral agents are specific to individual virus species.

The antiviral era

Antivirals generally act by targeting proteins that are unique to a virus and essential for its growth. Initially, these drugs were discovered by chance, often during the search for new cancer treatments, but now that many of the molecular pathways exploited by viruses have been unravelled, rational drug design can target individual viral proteins. However, with the ever-increasing antiviral drug repertoire, drug resistance has emerged as an increasing challenge due to viruses' high rate of replication, particularly with RNA viruses.

Acyclovir was the first effective antiviral available. It is active against many herpes viruses and is particularly useful for oral and genital herpes caused by herpes simplex virus as well as shingles caused by varicella zoster virus. The drug is quite safe and people at particular risk can take it over long periods to prevent unpleasant recurrences of these infections. Although long-term treatment can produce mutant herpes viruses that are resistant to acyclovir, fortunately so far the mutants have proved weaker than their non-mutated counterpart and have not spread in the general population.[1]

The next antivirals to be marketed targeted flu viruses. Despite being safe and effective, their use is largely restricted to those hospitalized with flu or with significant underlying health issues, because it is difficult to judge when and how to use them for milder cases. They can also be used to prevent flu, so during an epidemic they can be given to the elderly and chronically sick. Alternatively, they can be taken as soon as symptoms appear, when they may reduce the length of illness by about one day. This could have enormous financial benefit to employers by reducing absence from work, but is logistically challenging to deliver. Consequently, during the 2009 flu pandemic, many governments stockpiled drugs against flu, such as Tamiflu, but as the disease turned out to be milder than expected, they were rarely used. Vaccination remains the prevention of choice in high-risk populations, affording longer-term protection across various co-circulating seasonal strains.

ANTI-HIV-1 DRUGS

For at least a decade after the discovery of HIV in 1983, there was no really effective antiviral drug available for routine use against the virus, and the outlook for effective treatment or a cure for HIV infection was bleak. But by 1996, the situation had completely changed and optimism was so high that some predicted the imminent end of AIDS. The buzzword was 'triple therapy'. For the first time HIV-1 infection seemed to be treatable, and in the West death rates from AIDS began to fall. This remarkable revolution came about when AIDS doctors, taking a leaf out of the cancer doctors' book, began treating HIV patients with combinations of two or three antivirals at the same time, each with slightly different modes of action against the virus, rather than with a single drug.

As noted before, HIV-1 codes for three enzymes essential for its life cycle—protease, which ensures efficient infection and growth; reverse transcriptase (RT), which turns viral RNA into a DNA copy; and integrase, which ensures viral DNA integration into host cell chromosomes. These enzymes are targets for antiviral drugs that knock them out so that, even if eradication cannot be achieved, the virus can no longer propagate itself. More recently, drugs that inhibit HIV entry into cells by blocking its receptor complex have been added to our arsenal against this lethal virus.

The first drug treatment against HIV-1, zidovudine, also known as azidothymidine, was a failed cancer drug. It became available in 1987 and targets RT. Yet, although this was beneficial when first taken, drug-resistant mutants rapidly developed, rendering further treatment useless. But then as more antivirals were developed, combinations of drugs targeting different viral proteins efficiently inhibited viral replication and stabilized the infection. Therapeutic trials showed that a combination of three drugs, all taken together and including at least one protease inhibitor and one RT inhibitor, gave the best insurance against viral mutants appearing and causing the disease to progress. Such drug combinations became known as HAART (highly active anti-retroviral therapy) or, more recently, as just ART.

Monitoring the viral load in blood of an HIV-1-infected individual reflects how effectively the drugs are keeping the infection under control, and levels of blood CD4 helper T cells reflect how the immune system is coping. A rising viral load and falling CD4 counts indicates that it is time to change to another drug combination. ART halts progress of the infection and ensures the health of the immune system, so that most of those living with the virus can now expect a healthy life and normal lifespan. And although complete eradication is not possible, an undetectable viral load means that onward transmission is prevented.

Still, a small number of people living with HIV-1 do progress to AIDS. Often, these are patients unaware of their HIV-1 status, or individuals who present with the late complications triggering an HIV-1 test. However, patients with known HIV-1 and receiving ART do also sometimes progress to AIDS. This failure of control may be due to drug resistance, with people simply running out of drug options. But more commonly it is non-compliance with the drug regimen that is the problem. These HIV-1 carriers stop taking the drugs, often because of side effects or because they have so many tablets to take every day that it becomes impossible to keep up with every regimen. Until quite recently, the simplest drug regime for HIV-1 involved taking ten tablets a day, and for full triple therapy the number could reach 30, often taken at different times and frequencies. But this problem is being resolved as different drug combinations are processed into a single tablet that may eventually be taken just once daily.

Clearly, present ART is not ideal, but more long-acting drug combinations are being developed. One such is given monthly by injection, and this will certainly help with the compliance problem. Also, these drugs are now being used to prevent infection both before (pre-exposure prophylaxis) and after (post-exposure prophylaxis) potential exposure, with impressive effects. With these tools to hand, combined with increased HIV-1 testing to pick up silent infections, experts believe that the eventual end of HIV-1 transmission may now be in sight.

But all this only applies in affluent countries. The high cost and complexity of delivering ART and monitoring viral load have been constant barriers to widespread use in developing countries. However, for use across all populations in need, the WHO is now recommending simplified HIV treatment regimens prioritizing ART that is well tolerated, available in fixed dose combinations, and capable of being administered by primary care facilities. The WHO's global goal for the year 2020 was 90–90–90, meaning to diagnose 90 per cent of those living with HIV, to treat 90 per cent of cases, and to achieve viral suppression in 90 per cent of those treated. Unfortunately, the SARS-CoV-2 pandemic beginning in late 2019 intervened, and with healthcare workers diverted from HIV to SARS-CoV-2 control, these targets were missed. Nevertheless, HIV control has not been forgotten, and the new WHO global target is 95–95–95, to be hit by 2025.

ANTI-HEPATITIS B AND C DRUGS

The success of the HBV vaccine in preventing virus infection and spread has substantially reduced the global burden of disease, and as the vaccine becomes more widespread this reduction will continue. But there are still barriers to the overall success of the current regimen: up to 10 per cent of people do not appear to respond to the vaccine, rare vaccine escape mutant viruses are sometimes found in infants immunized at birth, and the need for multiple vaccine doses and the high cost limit its rollout.

Historically, medical therapy for those with chronic HBV infection aimed at enhancing the body's immune response to the virus, reducing viral replication in the liver and thereby preventing life-threatening complications such as cirrhosis and liver cancer. This generally involves administering a course of interferon alpha, a cytokine that spearheads the immune response. This is effective in 20–30 per cent of cases, but its use is limited by cost, length of treatment, and considerable side effects that some patients cannot tolerate. Better tolerated are novel antiviral drugs developed in the past ten years or so, which, by interrupting viral replication, reduce the incidence of liver failure and liver cancer, and these have become the standard of care. However, in contrast to interferon, which may result in virus clearance in a proportion of treated cases, they most often only control viral replication rather than clearing the infection, and so treatment may need to be lifelong.

In the converse situation to HBV, vaccine production against HCV has not been successful. This is mainly because HCV is an RNA virus with high turnover and a genome prone to mutations, and as such is likely to render any vaccine ineffective. Until recently, interferon treatment was used for persistent HCV infection, but, as with HBV treatment, patients suffered from side effects that were frequently intolerable. Fortunately, there has been a surge in production of direct-acting antiviral drugs against HCV taken by tablet and generally well tolerated (some of which are also active against HIV-1), and this has revolutionized HCV treatment. Unlike HBV, these agents are given in a time-limited course with the aim of cure, and when used in combination can achieve a response rate of over 95 per cent.

In this chapter we have followed the 200-year history of prevention of infectious diseases by vaccination, and discussed the more recent introduction of antiviral treatment strategies. Together these interventions have transformed infectious disease medicine and provided a healthier existence for all. The incredible success of the smallpox eradication programme demonstrated that humans can triumph over a lethal virus, and this success heralded a new era in the fight against infectious diseases. When it comes to elimination, each virus has its own particular defences to be overcome; some lurk in animal hosts, while some set up persistent infections which will be difficult to ferret out. For others we do not yet have a foolproof vaccine, but here the recent breakthroughs related to COVID-19 vaccine production provide new technologies that will surely break the deadlock. The long-term benefits of virus eradication, in financial as well as health terms, are clear to see, and although success may be several decades away it is certainly the way forward. In 2010, rinderpest, a relative of measles, became the second virus infection to be completely eradicated worldwide. Previously this virus caused cattle plague, a disease with a mortality rate approaching 100 per cent that destroyed millions of livelihoods, mainly in Africa and Asia. The WHO hit list of viruses ripe for eradication currently features polio and measles, now well on the way to extinction, while the campaigns against rabies and hepatitis B are at an earlier stage. COVID-19 must now be added to this list.

EPILOGUE: WHAT THE FUTURE HOLDS

In 2016, I wrote: 'remember, nasty surprises will continue to emerge—we must learn to expect the unexpected.' This was the last line of *Ebola: Profile of a Killer Virus*, a book about the 2014–16 Ebola outbreak in West Africa.[1] So have we remembered? Have we learnt lessons from each surprise epidemic or pandemic? Can we either prevent an outbreak occurring or respond before it goes global?

In relation to COVID-19, the answer to these questions was a resounding 'no'. The virus burst onto our radar screen in January 2020, and was unstoppable. Indeed, it was like watching a tsunami, with wave after wave engulfing towns, cities, and countries as it raced around the world causing a previously unknown lethal disease. The ever-rising infection and death rates left health services reeling. This was not only unforeseen but on a completely unprecedented scale in modern times.

For more than a year the world depended on nothing but medieval remedies like social distancing and lockdowns, and these came with their own severe problems—disrupted education, economic collapse, rising unemployment, increasing poverty, deteriorating mental health, and cancelled treatments for non-COVID-19-related illnesses—all accentuating social inequality.

Nevertheless, armed with our technological prowess and molecular know-how, the fightback was remarkably swift. If a virus can fly round the world at breakneck speed, then so can scientific information. Within weeks of Chinese scientists publishing SARS-CoV-2's genome sequence in January 2020, diagnostic tests and vaccines were in production and drug candidates were being sought. Just a year into the pandemic, life-saving vaccines were being rolled out around the world, and although there is a long way to go, hopefully the worst is over. Now we must learn to live with the consequences. All very impressive, but could we have stopped the pandemic before it even began?

The most recent pandemic of similar ferocity to COVID-19, the flu of 1918–19, came in three waves. So it is not surprising that this experience, combined with several more recent but less lethal flu pandemics, has shaped virtually all our pandemic preventative strategies. The WHO's Global Influenza Surveillance and Response System, mentioned in Chapter 8, is an enormous operation collecting data on emerging flu strains that are used to decide on the constituents of the annual vaccine. None of this is helpful in predicting or preventing any other emerging virus capable of epidemic and/or pandemic spread, but if we can do it for flu surely we can achieve the same for other emerging viruses.

Clearly emerging viruses are on the rise, so we urgently need to find out why they are emerging so frequently and how to stop them. We know that they are generally zoonotic, having jumped to us from an animal source, and broadly speaking the reason for the rise in these spillover events and subsequent spread is twofold: our burgeoning population, and increased international travel.

Human population growth

As far as population growth is concerned, we humans have become victims of our own success. In 1800 the global population, which had been growing steadily for centuries, hit one billion. It reached 1.6 billion by 1900, and now, in 2021, it stands at 7.8 billion (Figure 40). To put it bluntly, this population explosion is impacting every facet of life on Earth so that continued rise at this rate would be unsustainable for human existence as we know it. Overpopulation, combined with human greed, is the basis of a multitude of ills ranging from climate change to pandemics, and including poverty, water shortages, rainforest destruction, animal and plant extinctions, and pollution of land, air, and sea. By invading and destroying virgin territories in search of agricultural land, water, and mining or hunting opportunities, we inevitably disrupt natural ecosystems and displace wildlife. In doing so we cause long-lasting knock-on effects to indigenous plants, animals, and microbes,

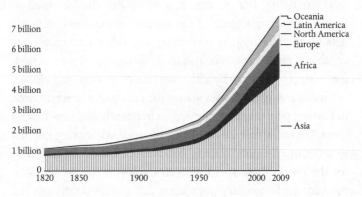

Figure 40. World population growth (1820–2009).

which viruses are quick to exploit. All too often the result of the imbalances we cause is spillover of a virus as it jumps species to cause a new human disease. These events are more common in tropical than in temperate climates because the huge diversity of animal and plant species in these areas supports a greater variety of microbes. What's more, recently spillover events have been increasingly reported in urban settings, suggesting the involvement of intermediate hosts bridging the gap between a virus reservoir in a wild animal and our urban populations.

A good example of human disturbance of nature causing a disease outbreak in animals with spillover to humans is the first emergence of Nipah virus in Malaysia in 1998. This began when farmers in Nipah and surrounding villages noticed an outbreak of respiratory disease among their pigs. The virus jumped from the pigs to the farmers and later to abattoir workers, causing encephalitis with a high mortality. The virus was dispersed further afield when infected pigs were moved, soon reaching neighbouring Singapore, but, fortunately, on this occasion the virus did not spread between humans. The outbreak was eventually controlled in 1999 by the slaughter of over a million pigs, but not before causing 265 human cases with 105 deaths.

Nipah virus was isolated from the brain of an encephalitis victim and traced to fruit bats. But the trail that led to the 1998 outbreak began much earlier, in the rainforests where deforestation was under way to clear virgin land for industrial plantations. This, along with drought conditions, caused a shortage of fruiting forest trees for foraging fruit bats, and led them to roost in the tempting mango orchards where farmed pigs also foraged. Virus in bat droppings was ingested by the pigs, causing the respiratory disease. Nipah virus is now known to be carried by at least 23 species of fruit bat that inhabit vast areas of forest in Madagascar, Cambodia, Indonesia, Thailand, the Philippines, and Ghana. The virus has re-emerged several times since 1999, particularly in India and Bangladesh, where human-to-human transmission has been recorded. Is this a disaster waiting to happen?[2]

International travel

Microbes have always exploited the human propensity to travel, be it for foraging, trading, discovery of new lands, war, famine, or, more recently, pleasure. Viruses that cause epidemics constantly require non-immune populations to infect, but epidemics can only move as fast as humans can travel. Yet, just as novel viruses are emerging at increasing rates, we have collapsed geographical space by speeding up international travel such that we can now reach the other side of the world in less than 24 hours. We are playing into their hands!

The travel time between the UK and Australia aptly illustrates my point. In the eighteenth century sailing ships took a year over the journey, while early nineteenth-century clippers took 100 days, and steam ships cut the journey time to 50 days at the beginning of the twentieth century. Despite these early, long and arduous journeys, viruses still travelled. Measles reached Australia around 1850, but on other routes viruses were much earlier colonizers. Yellow fever virus, for example, arrived in the Americas from Africa with transported slaves in the seventeenth century, while smallpox and other acute childhood viruses were carried to the Americas by the Spanish invaders in the sixteenth century.

The recent advent of rapid, frequent, affordable air travel entices 1.4 billion of us to travel internationally every year—a complete game changer for microbes. Diseases like HIV, SARS,

Ebola, flu, and now COVID-19 hopped between continents inside unsuspecting travellers often even before the threat was recognized. And there are many other less well-known examples, such as the following case of a virus carried by wild animals shipped between continents as part of the multimillion-dollar exotic pet trade.

The year 2003 saw a surprise outbreak of monkeypox in the US, the first time the disease had been diagnosed outside Africa, where it is endemic. Monkeypox virus is carried by rats (not monkeys), and is, as noted before, related to smallpox virus but causes a milder, non-fatal illness. The US outbreak involved five States, centred on the State of Wisconsin. Investigators pointed the finger at a consignment of wild animals shipped from Ghana, West Africa, to Texas. The shipment contained 800 Ghanaian mammals including various species of squirrels, rats, porcupines, and mice. Some of these animals ended up in a pet shop in Milwaukee, Illinois, including several giant pouched Gambian rats that were placed next to a cage of prairie dogs. It was the families that bought the prairie dogs that came down with monkeypox, the virus having jumped to the dogs from the cage of rats next door while the animals were housed in the pet shop. By the end of the outbreak, 49 cases of human monkeypox were identified, and the pet shop was prohibited from selling animals until the health of its animals could be verified.

Emerging virus surveillance

The WHO regularly publishes a list of potential emerging diseases that require urgent research and development. Presently this includes COVID-19, Ebola and related Marburg virus diseases, Lassa fever, MERS, SARS, Nipah and related henipaviral diseases, Rift Valley fever, Crimean–Congo haemorrhagic fever, Zika, and Disease X (meaning a hitherto unknown disease). But while most would agree that it is sensible to encourage research on these potential epidemic viruses, the most likely candidate to cause the next epidemic or pandemic is Virus X—a 'new' virus causing Disease X. In fact, in 2019 COVID-19 became Disease X caused by Virus X! So there is much debate among experts as to the best way to predict or prevent pandemics. For example, having noted the massive global effort that goes into flu surveillance, would it be possible to extend this to other potential emerging threats? Should we be proactive? Or reactive?

The proactivists are exemplified by The Global Virome Project (a privately funded programme), and PREDICT-2 (a USAID-funded programme which, after ten years of work, was cancelled by President Donald Trump just weeks before COVID-19 emerged). The aim of both programmes is to ensure 'the beginning of the end of the pandemic era' by identifying and characterizing novel viruses in wild mammals and birds that are a potential threat to humans. Obviously this is a mammoth task, requiring expertise in ecology, epidemiology, virology, cell and molecular biology, and genomics—but no one knows exactly how big the task is because we don't know how many viruses are out there. Some estimates suggest around 1.6 million, with perhaps 500,000 of them capable of infecting human cells and 10,000 of those being potential epidemic-causing viruses in humans; others think that there are fewer than 100 unknown viruses that pose a real threat to us. Whatever the exact figure, clearly identifying them all is a very long-term project. Indeed, in its decade of work PREDICT discovered more than 900 novel viruses, but just one had possible links to human infections. Nevertheless, PREDICT has worked to influence human behaviour in certain risky situations, such as handling animals in Asian

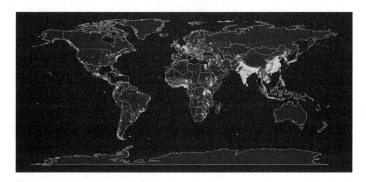

Figure 41. Map showing emerging infection hotspots.

wet markets where the SARS, and possibly the COVID-19, virus jumped to humans. This could be a more fruitful approach since along with the scientific input, pandemic prevention will require a regard for nature and a resolution to leave what is left of it untouched by human hands.

A map showing identified global hotspots for virus emergence is shown in Figure 41, but it is the scale of the problem that makes it difficult to envisage these projects fulfilling their goals any time soon. However, concentrating on bat viruses seems a sensible approach as these little mammals are highly mobile and are vectors for many recent zoonoses, including those causing Nipah, Ebola, SARS, MERS, and probably COVID-19. In fact, scientists in China at the Wuhan Institute of Virology, in the city where the COVID-19 pandemic began, have been studying bat viruses for several years. In 2013, the team isolated a bat coronavirus that grows in cultured cells from the human respiratory tract. This virus is 95 per cent identical to the present pandemic virus, but whether it could spill over to cause human disease remains a matter of speculation. Bats carry many different coronaviruses that have coevolved with them for millions of years—so how do we separate those that are potential threats to humans from those that are harmless?[3]

Scientists at the University of Edinburgh, UK, are attempting to answer this question. First, using statistical analysis of published data to narrow down the search for Virus X, they have produced the following identikit: a zoonotic, RNA virus from families, like the coronaviruses, that have previously produced viruses capable of causing epidemics. Now they are pinpointing the culprits further by using machine learning to predict which viruses are able to cause disease and spread among humans from detailed analysis of protein–protein interactions, particularly involving virus receptor molecules. And addressing the geographical location for potential spillover events, their published list of global hotspots prior to the current pandemic includes Wuhan, China. So this seems a fruitful line of research.[4]

Notwithstanding these findings, other scientists think that all this effort is of little value. They argue that the money would be better spent on boosting the capacity of health services around the world so that emerging infections can be spotted and dealt with before they spread. In this regard, the WHO was severely criticized for being unprepared and slow to recognize the extent of the problem in 2014–16, when Ebola unexpectedly ran riot in West Africa. Too little, too late was the verdict, and this cost lives.

In the years since 2016, the WHO has committed to strengthening national capacity for disease surveillance in low-income countries, so providing a strong global network, like flu surveillance, that reports infectious disease outbreaks from source to centre in time to curtail virus spread. This is a huge undertaking, and although we know what needs to be done, it is extremely expensive; it will only become a reality if governments around the world see it as a high priority.

In the year 2000, the Global Alliance Vaccine Initiative (GAVI) came into being. Its mission is to provide equal access to vaccines for children living in the poorest countries. Also, in 2017, in the wake of the 2014–16 Ebola epidemic and the frustration over Ebola vaccine production, which stalled for lack of funds, the Coalition for Epidemic Preparedness Innovations (CEPI) was launched. With the shared goals of preparing vaccines against emerging infections and enabling equitable access to them, these public–private partnerships have been enormously successful, and are needed now, during the COVID-19 pandemic, more than ever before. GAVI and CEPI, together with the WHO, are spearheading COVAX, 'the global initiative to ensure rapid and equitable access to COVID-19 vaccines to all countries, regardless of income level'.

Stopping pandemics such as COVID-19 depends on this initiative. In the case of SARS-CoV-2, we have already seen mutant viruses with increased transmissibility evolve and spread globally; until we can halt virus transmission everywhere, mutant viruses that are resistant to current vaccines may appear at any time.

Remember, no one is safe until everyone is safe.

REFERENCES

CHAPTER 1: THE VIROSPHERE

1. MEDAWAR, P. B. and MEDAWAR, J. S. (1983). *Aristotle to Zoos: A Philosophical Dictionary of Biology.* Harvard University Press, Cambridge, MA.
2. PAVORD, A. (1999). *The Tulip.* Bloomsbury Publishing, London.
3. Professor Norman Noah, Department of Public Health and Epidemiology, King's College London, Bessemer Road, London, SE5 9PJ. Personal communication.
4. ANDREWES, C., Sir (1973). *In Pursuit of the Common Cold.* William Heinemann Medical Books Limited, London.
5. BOSCH, X. (1998). 'Hepatitis C Outbreak Astounds Spain'. *Lancet* 351, 1415.
6. PETERSEN, L. R. and HAYES, E. B. (2004). 'Westward Ho?—The Spread of West Nile Fever'. New *England Journal of Medicine* 351, 2257–9.
7. MCALISTER GREGG, N. (1941). 'Congenital Cataract Following German Measles in the Mother'. *Transactions of the Ophthalmological Society of Australia,* 35–46.

Further reading

CRAWFORD, D. H. (2018). *VSI Viruses.* Second edition, Oxford University Press, Oxford, New York, Tokyo.
OXFORD, J., KELLAM, P., and COLLIER, L. (2016). *Human Virology.* Fifth edition, Oxford University Press, Oxford, New York, Tokyo.

CHAPTER 2: THE FIGHTBACK

1. ALVES, J. M., CARNEIRO, M., CHENG, J. Y., LEMOS DE MATOS, A., RAHMAN, M., LOOG, L., et al. (2019). 'Parallel Adaption of Rabbit Populations to Myxoma Virus'. *Science* 363, 1319–26.
2. ELSWORTH, P., COOKE, B. D., KOVALISKI, J., SINCLAIR, R., HOLMES, E. C., and STRIVE, T. (2014) 'Increased Virulence of

Rabbit Haemorrhagic Disease Virus Associated with Genetic Resistance in Australian Rabbits'. *Virology*, 464–5, 415–23.

3. MCNEIL, W. H. (1994). *Plagues and Peoples*. Penguin Books, London.

Further reading

KLENERMAN, P. (2018). *The Immune System*. Oxford University Press, Oxford, New York, Tokyo.

PLAYFAIR, J. and BANCROFT, G. (2013). *Immunity and Infection*. Oxford University Press, Oxford, New York, Tokyo.

CHAPTER 3: EMERGING INFECTIONS

1. WELLS, R. M., ESTANI, S. S., YADON, Z. E., et al. (1997). 'An Unusual Hantavirus Outbreak in Southern Argentina: Person-to-Person Transmission?' *Emerging Infectious Diseases*, 3:2, 171–4.

2. PIOT, P. (2012) *No Time to Lose*. W.W. Norton and Company, New York, London.

3. VARKEY, J. B., SHANTHA, J. G., CROZIER, I., et al. (2015). 'Persistence of Ebola Virus in Ocular Fluid during Convalescence'. *New England Journal of Medicine*, 372, 2423–7.

4. JACOBS, M., RODGERS, A., BELL, D., et al. (2016). 'A Case of Late Ebola Virus Relapse Causing Meningoencephalitis'. *Lancet*, 338, 498–503.

5. MATE, S. E., KUGELMAN, J. R., NYWNSWAH, T. G., et al. (2015). 'Molecular Evidence of Sexual Transmission of Ebola Virus'. *New England Journal of Medicine*, 373(25), 2448–54.

6. YU, I. T. S., LI, Y., WONG, T. W., et al. (2004). 'Evidence of Airborne Transmission of the Severe Acute Respiratory Syndrome Virus'. *New England Journal of Medicine*, 348, 1995–2005.

7. YU, Z. and SHI, Z. (2008). 'Investigation of Animal Reservoir(s) of SARS-CoV'. *Emerging Infections of Asia*. Springer Science+Business Media, LLC.

8. PARRY, J. (2004). 'Breaches of Safety are Probable Cause of Recent SARS Outbreak, WHO says'. *British Medical Journal*, 328(7450), 1222.

9. CHOI, S., JUNG, E., CHOI, B. Y., HUR, Y. J., and KI, M. (2018). 'High Reproduction Number of Middle East Respiratory Syndrome Coronavirus in Nosocomial Outbreaks: Mathematical Modelling in Saudi Arabia and South Korea'. *Journal of Hospital Medicine*, 99, 162–8.

Further reading

BARRETT, R. and ARMELAGOS, G. (2014). *An Unnatural History of Emerging Infections*. Oxford University Press, Oxford, New York, Tokyo.

CHAPTER 4: TWENTY-FIRST-CENTURY PANDEMICS

1. 'Pneumocystis Pneumonia—Los Angeles'. *The Morbidity and Mortality Weekly Report*, 30 (1981), no. 21.
2. 'Kaposi's Sarcoma and Pneumocystis Pneumonia among Homosexual Men'. *The Morbidity and Mortality Weekly Report*, 30 (1981), no. 25.
3. BARRÉ-SINOUSSI, F., CHERMANN, J. C., REY, E., et al. (1983). 'Isolation of a T-lymphotropic Retrovirus from a Patient at Risk for Acquired Immune Deficiency Syndrome (AIDS)'. *Science*, 220, 868–71.
4. JOHNSON, M. (1997). *Working a Miracle*. Bantam Books, New York, Toronto, London, Sydney, Auckland.
5. CREIGHTON, C. (1965). *History of Epidemics in Britain*. Vol. 2, (2nd edn), p. 308. Frank Cass, London.
6. SCHÄFER, W. (1955). 'Vergleichende sero-immunologische untersuchungen uber die viren der influenza and klassischen geflugelpest'. *Z. Naturfoschg* 10b, 81–91.
7. TAUBENBERGER, J. K. et al. (1997). 'Initial Genetic Characterization of the 1918 "Spanish" Influenza Virus'. *Science*, 275, 1793–6.
8. WOROBEY, M., PEKAR, J., LARSEN, B. B., et al. (2020). 'The Emergence of SARS-CoV-2 in Europe and North America'. *Science*, 370, 564–70.
9. LEMIEUX, J. E., SIDDLE, K. J., SHAW, B. E. et al. (2021). 'Phylogenetic Analysis of SARS-CoV-2 in Boston Highlights the Impact of Superspreading Events'. *Science*, 371, 588–97.
10. MALLAPATY, S. (2020) 'What the Cruise-Ship Outbreaks Reveal about COVID-19'. *Nature*, 580, 18.

Further reading

HOOPER, E. (2000). The *River: A Journey Back to the Source of HIV and AIDS*. Penguin, London and New York.
TAUBENBERGER, J. K., HULTIN, J. H., and MORENS, D. M. (2007). 'Discovery and Characterisation of the 1918 Pandemic Influenza Virus in Historical Context'. *Antiviral Therapy*, 12, 581–91.

CHAPTER 5: PAST EMERGING VIRUSES

1. BRIDGES, E. L. (1948). *Uttermost Part of the Earth.* Hodder and Stoughton, London.
2. 'Laboratory Work on Smallpox Virus'. *Lancet,* (i) (1979), 83–4.
3. FENNER, F., HENDERSON, D.A., ARITA, L., et al. (1988). 'Early Efforts at Control: Varioloation, Vaccination, and Isolation and Quarantine'. In: *Smallpox and Its Eradication.* World Health Organisation, Geneva, Switzerland.

CHAPTER 6: LIFELONG RESIDENTS

1. BARRÉ-SINOUSSI, F., CHERMANN, J. C., REY, F., et al (1983). 'Isolation of a T-Lymphotropic Retrovirus from a Patient at Risk for Acquired Immune Deficiency Syndrome (AIDS)'. *Science,* 220, 868–71.
2. GALLO, R. C., SALAHUDDIN, S. Z., POPOVIC, M., et al. (1984). 'Human T-lymphotropic Retrovirus, HTLV-III Isolated from AIDS Patients and Donors at Risk for AIDS'. *Science,* 224, 500–3.
3. DUESBERG, P. and ELLISON, B. J. (1996). *Inventing the AIDS Virus.* Regnery Publishing Inc., Washington, USA.
4. TOBLER, L. H. and BUSCH, M. P. (1997). 'History of Postperfusion Hepatitis'. *Clinical Chemistry,* 43:8(B), 1487–93.
5. BLUMBERG, B. S. (2002). *Hepatitis B: The Hunt for a Killer Virus.* Princeton University Press, Princeton, NJ, and Oxford.
6. BLUMBERG, B. S., GERSTLEY, B. J. S., HUNGERFORD, D. A., and LONDON, W. T. (1967). 'A Serum Antigen (Australia Antigen) in Down's Syndrome, Leukemia and Hepatitis'. *Annals of Internal Medicine,* 66(5), 924–31.
7. ZHAO, Z.-S., GRANUCCI, F., YEH, L., et al. (1998). 'Molecular Mimicry by Herpes Simplex Virus-Type 1: Autoimmune Disease after Viral Infection'. *Science,* 279, 1344–5.

CHAPTER 7: VIRUSES AND CANCER

1. HARRIS, H. (2004). 'Putting on the Brakes'. *Nature,* 427, 201.
2. BISHOP, J. M. (1990). Nobel Lecture. 'Retroviruses and Oncogenes II'. *Bioscience Reports,* 10, 473–91.

3. BURKITT, D. (1962). 'Determining the Climatic Limitations of a Children's Cancer Common in Africa'. *British Medical Journal*, 2, 1019–23.

4. EPSTEIN, M. A. Nuffield Department of Clinical Medicine, John Radcliffe Hospital, Headington, Oxford, OX3 9DU. Personal communication.

5. EPSTEIN, M. A., BARR, Y. M., and ACHONG, B. G. (1964). 'Virus Particles in Cultured Lymphoblasts from Burkitt's Lymphoma'. *Lancet*, (i), 702–3.

6. SHIBATA, D. et al. (1991). 'Association of Epstein-Barr Virus with Undifferentiated Gastric Carcinomas with Intense Lymphoid Infiltration. Lymphoepithelioma-like Carcinoma.' *American Journal of Pathology*, 139, 469–74.

7. WEISS, L. M. et al. (1987). 'Epstein-Barr Virus DNA in Tissues from Hodgkin's Disease'. *American Journal of Pathology*, 129, 86–91.

8. JONES, J. F. et al. (1988). 'T-cell Lymphomas Containing Epstein-Barr Viral DNA in Patients with Chronic Epstein-Barr Virus Infections'. *New England Journal of Medicine*, 318, 733–41.

9. CRAWFORD, D. H. et al. (1980). 'Epstein-Barr Virus Nuclear Antigen Positive Lymphoma after Cyclosporin A Treatment in Patient With Renal Allograft'. *Lancet*, 1, 1355–6.

10. ZUR HAUSEN, H. et al. (1970). 'EBV DNA in Biopsies of Burkitt Tumours and Anaplastic Carcinomas of the Nasopharynx'. *Nature*, 228, 1056–8.

11. POIESZ, B. J., RUSCETTI, F. W., GAZDAR, A. F., et al. (1980). 'Detection and Isolation of Type C Retrovirus Particles from Fresh and Cultured Lymphocytes of a Patient with Cutaneous T-cell Lymphoma'. *Proceedings of the National Academy of Sciences*, 77, no. 12, 7415–19.

12. HINUMA, Y., NAGATA, K., HANAOKA, M. et al. (1981). 'Adult T-cell Leukemia: Antigen in an ATL Cell Line and Detection of Antibodies to the Antigen in Human Sera'. *Proceedings of the National Academy of Sciences of the USA*, 78(10), 6476–80.

13. SCOTTO, J., BAILAR, J. C. (1969). 'Rigoni-Stern and Medical Statistics. A Nineteenth-Century Approach to Cancer Research'. *Journal of the History of Medicine and Allied Sciences*, 24, 65–75.

14. DURST, M., GISSMANN, L., IKENBERG, H., and ZUR HAUSEN, H. (1983). 'A Papillomavirus DNA from a Cervical Carcinoma and its Prevalence in Cancer Biopsy Samples from Different Geographical Regions'. *Proceedings of the National Acadademy of Sciences of the USA*, 80, 3812–15.

15. CHANG, Y., CESARMAN, E., and PESSIN, M. S., (1994). 'Identification of Herpesvirus-like DNA Sequences in AIDS-Associated Kaposi's Sarcoma'. *Science*, 265, 1865–9.

16. FENG, H., SHUDA, M., CHANG, Y., and MOORE, P. S. (2008). 'Clonal Integration of a Polyomavirus in Human Merkle Cell Carcinoma'. *Science*, 319, 1096–100.

17. MANOLOV, G. and MANOLOVA, Y. (1972). 'Marker Band in One Chromosome 14 from Burkitt Lymphomas'. *Nature*, 237, 33–4.

18. ROONEY, C. M., SMITH, C. A., NG, C. Y. C., et al. (1995). 'Use of gene-Modified Virus-Specific T Lymphocytes to Control Epstein–Barr Virus-Related Lymphoproliferation'. *Lancet*, 345, 9–12.

19. HAQUE, T., WILKIE, G. M., JONES, M. M., et al. (2007). 'Allogeneic Cytotoxic T Cell Therapy for EBV Positive Post-Transplant Lympho-proliferative Disease: Results of a Phase II Multicentre Clinical Trial'. *Blood*, 110, 1123–31.

Further reading

CHANG, Y., MOORE, P. S., and WEISS, R. A. (2017). 'Human Oncogenic Viruses: Nature and Discovery'. *Philosophical Transactions of the Royal Society B*, 372.

CRAWFORD, D. H., RICKINSON, A. B., and JOHANNESSEN, I. (2014). *Cancer Virus*. Oxford University Press, Oxford.

EPSTEIN, M. A. (1985). 'Historical Background: Burkitt's Lymphoma and Epstein–Barr Virus'. In: *Burkitt's Lymphoma: A Human Cancer Model* (ed. G. LENOIR, G. O'CONNOR, and C. L. M. OLWENY) 17–27. International Agency for Research on Cancer, Lyon (Scientific Publications, no. 60).

CHAPTER 8: TURNING THE TABLES

1. HALSBAND, R. (1953). 'New Light on Lady Mary Wortley Montagu's Contribution to Inoculation'. *Journal of the History of Medicine and Allied Sciences*, 8, no. 4, 390–405.

2. FENNER, F., HENDERSON, D. A., ARITA, L., et al. (1988). 'Early Efforts at Control: Varioloation, Vaccination, and Isolation and Quarantine'. In: *Smallpox and Its Eradication*. World Health Organization, Geneva, p. 260.

3. WATERSON, A. P. and WILKINSON, L. (1978). 'Early Terminology and Underlying Ideas'. In: *An Introduction to the History of Virology*. Cambridge University Press, Cambridge, p. 6.

4. MCNEILL, W. H. (1976). 'The Ecological Impact of Medical Science and Organisation Since 1700'. In: *Plagues and Peoples*. Penguin Books, London, p. 231.

5. WAKEFIELD, A. J., MURCH, S. H., ANTHONY, A., et al. (1998). 'Ileallymphoid-Nodular Hyperplasia, Non-Specific Colitis, and Pervasive Developmental Disorder in Children'. *Lancet,* 351, 637–41.

EPILOGUE: WHAT THE FUTURE HOLDS

1. CRAWFORD, D. H. (2016). *Ebola: Profile of a Killer Virus*. Oxford University Press, Oxford.

2. EPSTEIN, J. H., ANTHONY, S. J., ISLAM, A. et al. (2020). 'Nipah Virus Dynamics in Bats and Implications for Spillover to Humans'. *Proceedings of the National Academy of Sciences of the USA,* 117(46), 29190–201.

3. MENACHERY, V. D., YOUNT, B. L., DEBBINK, K. et al. (2015). 'A SARS-like Cluster of Circulating Bat Coronaviruses Shows Potential for Human Emergence'. *Nature Medicine,* 21, 1508–13.

4. WOOLHOUSE, M. E. J. and ASHWORTH, J. L. (2017). 'Can We Identify Viruses with Pandemic Potential?' *The Biochemist* (London), 39:3, 8–11.
WOOLHOUSE, M. E. J., BRIERLEY, L., MCCAFFERY, C., and LYCELL, S. (2016) 'Assessing the Epidemic Potential of RNA and DNA Viruses'. *Emerging Infectious Diseases,* 22, 2037–44.
ZHANG, F., CHASE-TOPPING, M., GUO, C.-G. et al. (2020). 'Global Discovery of Human-Infective RNA Viruses: A Modelling Analysis'. *PLOS Pathogens,* November 30.

GLOSSARY OF TERMS

acquired immunodeficiency syndrome (AIDS): an immune deficiency caused by infection with human immunodeficiency virus, characterized by opportunistic infections.

angiotensin converting enzyme-2: a cell surface protein, widespread on cells in the lungs, arteries, heart, kidneys, and intestine, which acts as the receptor for SARS-CoV and SARS-CoV-2.

acyclovir: a drug which inhibits the growth of certain herpes viruses. Used mostly to treat or prevent genital and oral herpes and shingles.

adenovirus: DNA virus which takes its name from the adenoid—the human tissue in which it was first found. Causes respiratory tract and eye infections.

amino acid: a simple organic compound naturally occurring in plant and animal tissues and forming the basic constituent of proteins.

anthrax: a disease caused by infection with *Bacillus anthracis* from infected animals. Causes haemorrhage and effusions in various organs and is often fatal.

antibiotic: a substance which can inhibit or destroy susceptible micro-organisms. (From *anti-bios*, Greek for 'against life'.)

antibody: a blood protein capable of binding to and neutralizing an antigen.

antigen: a foreign substance, usually a protein constituent of a microbe, capable of inducing an immune response in the body.

antigenic drift: accumulated mutations in flu virus genetic material which eventually differs from the parent virus enough to cause an epidemic.

antigenic shift: a major genetic change in flu virus arising by genetic reasssortment which may cause a flu pandemic.

aphid: greenfly or blackfly, an insect that sucks the sap from plants.

aphrodisiac: a drug which arouses sexual desire. (Named after Aphrodite, the Greek goddess of love.)

apoptosis: controlled death of cells. (From the Greek *apo* and *ptosis*, meaning 'falling off'.) Also called programmed cell death or the suicide programme.

arbovirus: a large group of RNA viruses, which are spread by arthropod vectors. (Name derived from arthropod-borne viruses).

attenuate: reduce in virulence. For a virus, this is usually by prolonged culture under unfavourable conditions.

Australia antigen: a protein found in the blood of hepatitis B virus carriers, now known to be HBV surface antigen.

autoimmunity: a condition caused by antibodies and/or reactive lymphocytes directed against normal body substances. Causes autoimmune disease.

B cell: *see* B lymphocyte.

bacterium: a unicellular microorganism.

bilharzia: a parasitic worm infection also known as schistosomiasis.

botulism: a fatal illness caused by the toxin of the bacterium *Clostridium botulinum*.

BRCA 1, 2: breast cancer gene 1 and 2, genes that raise the risk of breast and ovarian cancer.

bubonic plague: An infectious disease caused by the bacterium *Yersinia pestis*.

Canine distemper virus: a paramyxovirus related to measles virus that infects many mammal species including dogs.

capsid: the outer protective coat of a virus.

CCR5: chemokine receptor type 5, part of the cell receptor for HIV.

CD4: a marker denoting a 'helper' T lymphocyte. The type of cell susceptible to HIV infection.

cell cycle: the reproductive process of a cell resulting in cell division.

central nervous system: the nervous tissue of the brain and spinal cord.

cervix: the neck of the womb or uterus. (From the Latin 'cervix' meaning 'neck'.)

cirrhosis: a chronic disease of the liver in which glandular tissue is replaced by fibrous scar tissue. May be caused by alcoholism or hepatitis B or C viruses. (Named by Laennec from the Greek *Kirrhos* meaning 'tawny'.)

chlamydia: A sexually transmitted bacterium.

c-myc: a cellular oncogene that is translocated in Burkitt lymphoma cells.

congenital: existing from birth.

cornea: clear layer of tissue in front of the eye through which light passes.

coronaviruses: a family of viruses named after the crown-like appearance of the virus particle (corona is derived from the Latin for crown). The family includes four common cold viruses and SARS, MERS, and SARS-2 coronaviruses.

cytokine: a small peptide molecule which initiates an immune response. Examples include the interferons and interleukins.

cytokine storm: An unregulated outpouring of cytokines.

cytomegalovirus (CMV): a persistent herpes virus. Infection in utero may cause cytomegalic disease of the newborn, and infection later in life can cause a glandular fever-like illness. In the immunocompromised host, CMV can cause pneumonitis, colitis, and retinitis leading to blindness. (Named after the swollen appearance of infected cells: cytomegalo = large cell.)

cytoplasm: the contents of a cell confined by the cell wall and surrounding the nucleus.

dengue fever: a flu-like illness consisting of fever, joint and muscle pains, and rash. Caused by the arbovirus, dengue fever virus. Spread by the mosquito *Aedes aegypti*. The virus can also cause the more severe dengue haemorrhagic fever. (The name is derived from Swahili 'denga', assimilated into West Indian Spanish and meaning 'fastidiousness', referring to the stiffness of a patient's neck and shoulders.)

DNA: deoxyribonucleic acid. The carrier of genetic information and the constituent of chromosomes. Present in nearly all living cells.

Ebola disease virus: a filovirus which causes severe and often fatal haemorrhagic fever. Epidemic disease occurs in equatorial Africa. (Named after the river in Zaire where the first recorded outbreak of Ebola fever happened.)

electron microscope: a microscope which uses a beam of electrons instead of light. Magnifies up to 100,000 times.

encephalitis: inflammation of the brain.

encepthalomyelitis: inflammation of the brain and spinal cord.

endemic: present in a community or among a group of people. An endemic disease is present continually in a particular region.

enzyme: protein produced by living cells which catalyses specific biochemical reactions.

epidemic: an unusual increase in the number of cases of a disease in a community.

epidemiology: the study of disease and disease attributes in defined populations. It is the scientific basis for public health and preventive medicine.

epithelium: the layer of cells covering an internal body surface such as the intestinal and respiratory tracts.

Epstein–Barr virus: a herpes virus which is the cause of glandular fever and is associated with a variety of human tumours. (Named after the two discoverers of the virus in 1964.)

filtrate: the liquid that passes through a filter.

gene reassortment: gene swapping in flu viruses which causes an antigenic shift.

genome: the full set of chromosomes of a cell or organism.

glandular fever: an acute disease caused by primary Epstein–Barr virus infection. Also called infectious mononucleosis.

gp120: an HIV-coded protein of molecular weight 120 kilodaltons. Acts as the viral receptor by binding to the CD4 molecule on susceptible cells.

grey matter: Part of the brain and spinal cord containing nerve cell bodies, and grey in colour.

growth factors: soluble proteins, produced by cells, which stimulate the growth of other cells.

haemagglutinin: a molecule on the surface of flu virus which binds red blood cells.

haemophilia: an inherited lack of blood clotting factor VIII which leads to a tendency to excessive bleeding when a blood vessel is injured.

haemorrhagic fever: a syndrome caused by a variety of viruses, including yellow fever, dengue, and Ebola viruses. Causes damage to blood vessels and bleeding which is often fatal.

hantavirus: a Bunyavirus (named after the place where the first member of the family was isolated—Bunyamwera, Uganda) which is transmitted to humans by rodents. Causes hantavirus pulmonary syndrome. (Named after the Hantaan river in Korea where the virus was first identified.)

hepatitis: inflammation of the liver.

hepatitis A virus: a picorna virus which causes acute viral hepatitis and is transmitted by the faecal–oral route.

hepatitis B virus: a hepadnavirus which causes serum hepatitis. Infection may lead to chronic hepatitis, cirrhosis, or liver cancer. Usually transmitted by injection of infected blood or by use of contaminated needles. (Name derived from hepa (meaning liver)-DNA-virus).

hepatitis C virus: a flavivirus which is the principal cause of non-A and non-B post-transfusion hepatitis.

herd immunity: The level of immunity within a community that protects against a particular infection.

herpes stromal keratitis: an autoimmune disease of the eye caused by herpes simplex virus.

herpesviruses: a family of DNA viruses. There are eight human herpesviruses including those causing cold sores, chickenpox, shingles, and glandular fever. (The name herpes is derived from the Greek *herpeton* meaning 'reptile' and refers to the creeping nature of the lesions—probably shingles.)

HHV 6 and HHV 7: Human herpes viruses 6 and 7. Infection is widespread in the human population but with no known disease associations.

holoendemic malaria: malaria caused by *Plasmodium falciparum* that occurs at the same intensity throughout the year.

human immunodeficiency virus (HIV): a retrovirus named after the immune deficiency it causes which leads to AIDS. There are two types: HIV-1 and HIV-2.

hydrophobia: a fear of drinking, typical of rabies infection.

immunological memory: the ability to prevent disease on second infection with a microorganism due to T- and B-cell memory.

index case: the first identified case in an outbreak, from which all other cases are derived.

influenza virus: an orthomyxovirus which causes flu epidemics and pandemics.

interferon: one of the cytokines that initiate an immune response.

interleukin-2: a cytokine that stimulates T-cell growth.

Kaposi's sarcoma: a tumour of the cells which line blood vessels (endothelial cells). Common in AIDS patients, it causes red patches on the skin. It is associated with infection with human herpesvirus 8.

Lassa fever virus: an arenavirus which causes Lassa fever. (Named after the village in Nigeria where an outbreak occurred which resulted in the isolation of the virus.)

leukaemia: a malignant disease of the white corpuscles in the blood.

lymph glands/nodes: a small mass of tissue where lymphocytes congregate and multiply.

lymphocyte-B: the type of white blood cell which produces antibodies.

lymphocyte-T: the type of white blood cell mainly responsible for immunity to viruses. Includes CD4 'helper' T cells and CD8 'killer' T cells.

lymphoma: a tumour of lymphoid tissue, consisting of malignant B or T cells.

lyssavirus: a genus of viruses including the rabies virus. (Name derived from the Greek *Lyssa* meaning madness).

macrophage: a bone marrow-derived cell which engulfs foreign or dead material in tissues.

mimivirus: a giant virus found in amoebae.

mosquito: A blood-sucking fly with aquatic lava. *Aedes aegypti* mosquitoes spread yellow fever, Zika, and dengue viruses.

mucous membrane: the general name given to the membrane which lines many of the hollow organs of the body.

multiple sclerosis (MS): a disease of the brain and spinal cord. Over a period of time it can cause paralysis and tremors.

mutation: an instant genetic change or alteration which, when transmitted to offspring, gives rise to inheritable variations.

myelin basic protein: a major protein constituent of nerve tissue.

myxomatosis: a fatal disease of European rabbits marked by conjunctivitis and the development of myxomatous growths in the skin; caused by rabbit myxoma virus.

neuraminidase: an enzyme on the surface of flu viruses which can destroy neuraminic (sialic) acid.

neutralizing antibody: an antibody that prevents infection of a cell by binding to a virus's receptor molecule.

nucleus: the central body in a cell, which contains the chromosomes.

obligate parasite: a parasitic organism, such as a virus, that cannot complete its life cycle without a host.

oncogene: a cellular gene that stimulates cell growth. Viral oncogenes can cause unregulated cell growth and induce tumour formation (oncogenesis).

ophthalmic nerve: a branch of the trigeminal nerve that supplies the eye.

opportunistic infection or tumour: an infectious disease or virus-associated tumour that occurs when the host's immunity is suppressed.

pandemic: an epidemic spreading on more than one continent at once.

panzootic: a pandemic among animals.

papillomaviruses: a genus of viruses which cause benign warts and malignant epithelial tumours of uterine cervix, penis, and larynx. (Named from Latin *papilla* meaning 'nipple'.)

pathogen: any microorganism causing disease.

peptide: a compound formed by the union of two or more amino acids which can itself form part of a protein.

perinatal: relating to the time immediately before and after birth.

placebo: a medicine (or tablet) given which will not have any definite action.

placenta: the afterbirth.

plague: the name of an infectious epidemic disease caused by the bacterium *Yersinia pestis*. Common in the Middle Ages. Causes fever and swelling of the lymphatic glands and carries a very high mortality. Rats are the natural reservoir.

plasmid: DNA (chiefly bacterial) which is separate from the chromosome of the host cell and can replicate but is not essential for the cell's survival.

Pneumocystis jirovecii: a single-celled organism related to fungi. Causes pneumonia in the immunocompromised host.

pneumonitis: inflammation of the lungs that does not progress to pneumonia.

polio virus: an enterovirus which causes poliomyelitis. Derived from the Greek *polios* meaning 'grey' and *myelos* meaning 'marrow'.

polyoma virus: a virus of the papovavirus family, which cause tumours in animals. (Name derived from the Greek *poly* meaning many, *oma* meaning tumours).

poxviruses: a family of viruses including the cause of smallpox. (Derived from the Anglo-Saxon word *pokkes* meaning 'pouch' and referring to the skin lesions.)

primate: the highest order of mammals, which includes tarsiers, lemurs, apes, monkeys, and humans.

programmed cell death: another term for apoptosis.

protease: an enzyme which can break down proteins and peptides; a proteolytic enzyme.

protein: nitrogenous substance widely distributed in the animal and vegetable kingdoms and forming the characteristic materials of tissues and fluids. Essentially combinations of amino acids.

protozoa: free-living, mobile, unicellular organisms, including amoebae (which can induce dysentery) and plasmodia (the cause of malaria).

rabbit haemorrhagic fever virus: a calcivirus which causes severe haemorrhagic fever in susceptible rabbits.

rabies: a fatal disease which infects dogs, wolves, and other mammals. Infection is from the saliva of a rabid animal. Caused by a virus of the rhabdovirus family (derived from the Greek *rhabdos* meaning 'rod').

receptor: a molecule in a cell membrane which binds specifically to a substance.

reproductive number (R): the number of new cases derived from a single case during an epidemic or pandemic. Ro is the value of R at the beginning of an epidemic and is constant for a particular virus or virus strain.

retinitis: inflammation of the retina.

retroviruses: a group of RNA viruses which insert a DNA copy of their genome into host chromosomes in order to persist.

reverse transcriptase: an enzyme which catalyses the conversion of RNA to DNA. Carried by retroviruses and required for their integration into the host cell genome.

rhinovirus: a picornavirus which causes the common cold. (The name is derived from the Greek *rhis* meaning 'nose'.)

rinderpest virus: virus of the genus *Morbillivirus*, causing the acute, highly contagious disease Rinderpest in ruminants and pigs.

RNA: ribonucleic acid. One of the two types of nucleic acid that exist in nature (the other is DNA). It is present in both the cytoplasm and nucleus of cells.

Rotavirus: a group of viruses so called because of their wheel-like structure. A common cause of gastroenteritis in infants. (The name derives from the Latin *rota* for 'wheel'.)

rubella: another name for German measles (*rubellus* being Latin for 'reddish' and referring to the symptomatic rash). Caused by rubella virus, a togavirus (the name being derived from the toga-like, close-fitting envelope surrounding the virus).

sialic acids: cell-surface glycoproteins and glycolipids that act as the cell receptor for flu viruses, binding to the viral haemagglutinin molecule.

simian immunodeficiency virus (SIV): a retrovirus which causes an AIDS-like syndrome in certain species of old-world primates.

sin nombre virus: a type of hantavirus. (Derived from the Spanish for 'without name'.)

slim disease: a wasting disease commonly seen in African AIDS.

stem cell: an undifferentiated cell from which specialized cells develop.

sterile immunity: immunity produced by natural infection or vaccination that prevents infection as well as disease.

stromal herpes keratitis: herpes virus infection of the eye which can cause scarring, leading to blindness.

subunit vaccine: vaccine made from parts of viruses, such as individual proteins, which induce an immune response.

T cell: *see* T lymphocyte.

tissue culture: the growth in an artificial medium of cells derived from living tissue.

transposon: a DNA sequence that can move around the genome, also called a jumping gene.

Treponema pallidum: a spirochaete bacterium that causes syphilis.

trigeminal nerve: the fifth cranial nerve, which supplies the skin of the face. It has three divisions: the ophthalmic, maxillary, and mandibular nerves.

tropical spastic paraparesis: a chronic neurological disease which occurs in the Caribbean and Japan and is associated with human T-cell leukaemia virus infection.

tumour suppressor genes: genes which negatively control cell division. Mutations of these genes are associated with many types of cancer. Also called anti-oncogenes.

vaccination: the process of inoculating with a vaccine to obtain immunity, or protection, against the corresponding disease. (Derived from the Latin *vacca* for 'cow'.)

vaccine: a substance, either dead or attenuated living infectious material, introduced into the body to produce resistance to a disease.

vaccinia virus: a pox virus used in vaccination against smallpox. The origins of vaccinia virus are obscure.

varicella: another name for chickenpox.

variolation: an early form of inoculation against smallpox.

verruca: a wart on the sole of the foot.

virosphere: the world of viruses.

virus particle: the extracellular form of a virus consisting of a protein coat or capsid surrounding the genetic material (DNA or RNA).

wet market: a market where animals are sold live for the table. Also called live animal markets, they are common in the Far East.

yellow fever: an acute viral infection of certain tropical localities. Characterized by fever and jaundice.

zidovudine (AZT): a derivative of thymine used to treat HIV and other viral infections.

zoster: shingles. A skin eruption, caused by varicella zoster virus, consisting of small yellow vesicles. (From the Greek word *circingle* meaning 'girdle', because it spreads in a zone-like manner round half the chest.)

zoonosis: a disease of humans acquired from an animal source.

INDEX

Note: f and *t* after page numbers indicate figures and tables.

Titles in the *Oxford Landmark Science* series